A FINITE ELEMENT METHOD PRIMER

FOR MECHANICAL DESIGN

The PWS Series in Engineering

Anderson, *Thermodynamics*
Askeland, *The Science and Engineering of Materials*, Third Edition
Bolluyt/Stewart/Oladipupo, *Modeling for Design using SilverScreen*
Borse, *FORTRAN 77 and Numerical Methods for Engineers*, Second Edition
Clements, *68000 Family Assembly Language*
Clements, *Microprocessor Systems Design*, Second Edition
Clements, *Principles of Computer Hardware*, Second Edition
Das, *Principles of Foundation Engineering*, Second Edition
Das, *Principles of Geotechnical Engineering*, Third Edition
Das, *Principles of Soil Dynamics*
El Wakil, *Materials Science and Engineering Lab Manual*
Fleischer, *Introduction to Engineering Economy*
Gere/Timoshenko, *Mechanics of Materials*, Third Edition
Glover/Sarma, *Power System Analysis and Design*, Second Edition
Janna, *Design of Fluid Thermal Systems*
Janna, *Introduction to Fluid Mechanics*, Third Edition
Kassimali, *Structural Analysis*
Keedy, *Introduction to CAD using CADKEY*, Third Edition
Keedy/Teske, *Engineering Design Graphics Using CADKEY*
Knight, *A Finite Element Method Primer for Mechanical Design*
Knight, *The Finite Element Method in Mechanical Design*
Logan, *A First Course in the Finite Element Method*, Second Edition
McGill/King, *Engineering Mechanics: Statics*, Second Edition
McGill/King, *Engineering Mechanics: An Introduction to Dynamics*, Second Edition
McGill/King, *Engineering Mechanics: Statics and An Introduction to Dynamics*, Second Edition
Raines, *Software for Mechanics of Materials*
Reed-Hill/Abbaschian, *Physical Metallurgy Principles*, Third Edition
Reynolds, *Unit Operations and Processes in Environmental Engineering*
Sack, *Matrix Structural Analysis*
Schmidt/Wong, *Fundamentals of Surveying*, Third Edition
Segui, *Fundamentals of Structural Steel Design*
Segui, *LRFD Steel Design*
Shen/Kong, *Applied Electromagnetism*, Second Edition
Stewart/Bolluyt/Oladipupo, *Modeling for Design using AutoCAD AME and AutoSurf*
Strum/Kirk, *Contemporary Linear Systems using MATLAB*
Sule, *Manufacturing Facilities: Location, Planning, and Design*, Second Edition
Vardeman, *Statistics for Engineering Problem Solving*
Weinman, *VAX FORTRAN*, Second Edition
Weinman, *FORTRAN for Scientists and Engineers*
Zirkel/Berlinger, *Understanding FORTRAN 77 and 90*

A FINITE ELEMENT METHOD PRIMER
FOR MECHANICAL DESIGN

Charles E. Knight, Jr.
Department of Mechanical Engineering
Virginia Polytechnic Institute
and State University

PWS Publishing Company
Boston

PWS PUBLISHING COMPANY
20 Park Plaza, Boston, MA 02116-4324

PWS Publishing Company is a division of Wadsworth, Inc.

I(T)P ™
International Thomson Publishing
The trademark ITP is used under license

Library of Congress Cataloging-in-Publication Data

Knight, Charles E.
A finite element method primer for mechanical design / Charles E. Knight, Jr.
p. cm.
Includes bibliographical references and index.
ISBN 0-534-93978-3
1. Finite element method. 2. Engineering design--Data processing. 3. Computer-aided design. I. Title.
TA347.F5K66 1993 93-27512
620'.0042'0151535--dc20 CIP

Sponsoring Editor: Jonathan Plant
Assistant Editor: Mary Thomas
Editorial Assistant: Cynthia Harris
Production Editor: Monique Calello
Marketing Manager: Nathan Wilbur
Cover Designer: Monique Calello
Manufacturing Manager: Ellen Glisker
Text Printer and Binder: Malloy Lithographing

Printed and bound in the United States of America.
93 94 95 96 97 98 -- 10 9 8 7 6 5 4 3 2 1

This book is dedicated to my son and daughter, Michael and Laura

PREFACE

A *Finite Element Method Primer for Mechanical Design* was written to be a supplementary text for the junior mechanical design course in a traditional mechanical engineering curriculum. The intent is to introduce the finite element method in the context of mechanical design. Emphasis is on the practical aspects of proper modeling, checking, and interpretation of results. Theoretical aspects are introduced as they are needed to help understand the operation. The text covers truss, beam/frame, and two dimensional solid engineering structures. This text is basically an excerpt from *The Finite Element Method in Mechanical Design*.

The finite element method has rapidly become a vital tool for analysis of mechanical designs. It has reached the point where practically every design engineer has access to a finite element program through their company's mainframe computer or on a micro-computer. This method is extremely powerful in terms of the many different types of problems it can solve. The method applies to many engineering fields, however this book concentrates on the application to mechanical design.

Use of this tool does not guarantee correct results. It is a numerical procedure involving approximations of theoretical behavior. In order to produce correct results the model must be designed correctly, and be able to reach numerical convergence (assuming the computer program is without error). Correct results then depend primarily on the user's ability to utilize the tool. It in no way supplants the engineer's responsibility to do approximate engineering calculations, use good design practice, and apply engineering judgment to the problem. Instead, it should supplement these skills to ensure that the best design is obtained.

With the ready access of finite element analysis comes the need to provide the understanding required to accurately and effectively use the method. Most design engineers are not going to have the time nor the inclination to study all the theoretical formulations and computer algorithms much less write their own program. Therefore, most engineers take the role of users of an in-house or commercial program.

Information About Accompanying Disk

A PC computer disk is provided with this book that contains a finite element program called FEPC. The program runs on an IBM PC or compatible computer. It will solve two-dimensional problems using truss, beam, plane stress, plane strain or axisymmetric solid elements. Dimensional limits of the program are described in an abbreviated user's guide in the Appendix. Also, you should review the README file on the disk for start-up instructions. The complete user's guide is in a file on the disk. The program is provided as shareware.

Although technical support can not be provided, we will replace any defective disk. Also, you may contact the author regarding any program errors or availability of program updates.

Acknowledgments

I would like to thank my former students in the course for their suggestions in refining the material for this text. Thanks to my colleagues and department head in the Department of Mechanical Engineering for their encouragement and support. My appreciation goes to Jonathan Plant, Monique Calello and the staff at PWS Publishing Company for their work, understanding and patience throughout this project. Finally, thanks to the following individuals who reviewed this text and provided many useful comments and suggestions that helped to improve the final product: A. Henry Hagerdoorn, *University of Central Florida*; H. Kazerooni, *University of California - Berkeley*; Julius Wong, *University of Louisville*; and Larry Stauffer, *University of Idaho*.

Charles E. Knight, Jr.

CONTENTS

CHAPTER 4

TWO-DIMENSIONAL SOLIDS 40

APPENDIX 58

CHAPTER 1

THE FINITE ELEMENT METHOD

1.1 General Overview

The finite element method is enjoying widespread use in many engineering applications. Although first developed for structural analysis, it now solves problems in heat transfer, fluid mechanics, acoustics, electromagnetics, and other specialized disciplines. In conduction heat transfer, we solve for the temperature distribution throughout the body with known boundary conditions and material properties whether steady state or time dependent. Application to fluid mechanics begins with steady inviscid incompressible flow and progresses to very complex viscous compressible flow. The whole area of computational fluid dynamics has made rapid progress in recent years. Acoustics is another area where great strides are being made based on finite element and boundary element numerical methods. Electromagnetic solutions for magnetic field strength provide insight for design of electromagnetic devices. Many of these capabilities are now being coupled to yield solutions to fluid-structure interactions, convective heat transfer, and other coupled problems.

The finite element method is a numerical method for solving a system of governing equations over the domain of a continuous physical system. The method applies to many fields of science and engineering, but this text focuses on its application to structural analysis. The field of continuum mechanics and theory of elasticity provide the governing equations.

The basis of the finite element method for analysis of solid structures is summarized in the following steps. Small parts called *elements* subdivide the domain of the solid structure illustrated in Figure 1-1. These elements assemble through interconnection at a finite number of points on each element called *nodes*. This assembly provides a model of the solid structure. Within the domain of each element we assume a simple general solution to the governing equations. The specific solution for each element becomes a function of unknown solution values at the nodes. Application of the general solution form to all the elements results in a finite set of algebraic equations to be solved for the unknown nodal values. By subdividing a structure in this manner, one can formulate equations for each separate finite element which are then combined to obtain the solution of the whole physical system. If the structure response is linear elastic, the algebraic equations are linear and are solved with common numerical procedures.

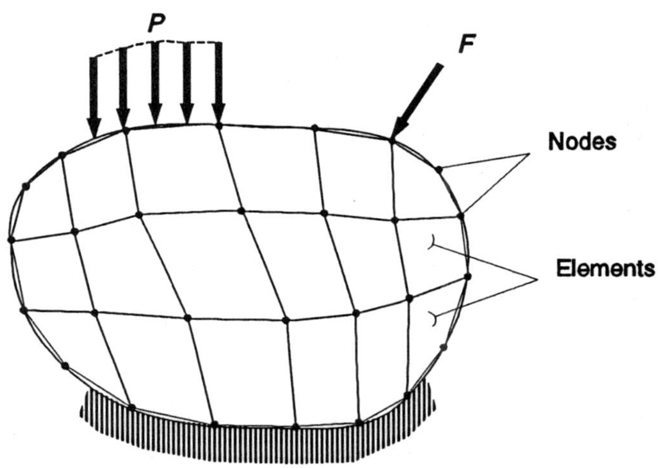

Figure 1-1. Two-Dimensional Continuum Domain

Since the continuum domain is divided into finite elements with nodal values as solution unknowns, the structure loads and displacement boundary conditions must translate to nodal quantities. Single forces like *F* apply to nodes directly while distributed loads like *P* are converted to equivalent nodal values. Supports like the grounding indicated by the hatch in Figure 1-1 resolve into specified displacements for the supported nodes.

At least two sources of error are now apparent. The assumed solution within the element is rarely the exact solution. The error is the difference between assumed and exact solutions. The magnitude of this error depends on the size of the elements in the subdivision relative to the solution variation. Fortunately, most element formulations converge to the correct solution as the element size reduces. The second error source is the precision of the algebraic equation solution. This is a function of the computer accuracy, the computer algorithm, the number of equations, and the element subdivision. Both error sources are reduced with good modeling practices.

In theory all solid structures could be modeled with three-dimensional solid continuum elements. However, this is impractical since many structures are simplified with correct assumptions without any loss of accuracy, and to do so greatly reduces the effort required to reach a solution. Different types of elements are formulated to address each class of structure. Elements are broadly grouped into two categories, structural elements and continuum elements.

Structural elements are trusses, beams, plates, and shells. Their formulation uses the same general assumptions about behavior as in their respective structural theories. Finite element solutions using structural elements are then no more accurate than a valid solution using conventional beam or plate theory, for example. However, it is usually far easier to get a finite element solution for a beam, plate, or shell problem than it is using conventional theory.

Continuum elements are the two- and three-dimensional solid elements. Their formulation basis comes from the theory of elasticity. The theory of elasticity provides the governing equations for the deformation and stress response of a linear elastic continuum subjected to external loads. Few closed form or numerical solutions exist for two-dimensional continuum problems, and almost none exist for three-dimensional problems; this makes the finite element method invaluable.

An extensive literature has developed since the 1960s when the term "finite element" originated. The first textbook appeared in 1967 [see Reference 1.1]. The number of books and conference proceedings published since then is near two hundred and the number of journal papers and other publications is in the thousands. The engineer beginning study of the finite element method may consult references [1.2], [1.3], [1.4], [1.5], [1.6], [1.7], or [1.8] for formulation fundamentals, [1.9], [1.10], [1.11], [1.12], [1.13], or [1.14] for structural and solid mechanics applications, and [1.15], [1.16], or [1.17] for computer algorithms and implementation.

1.2 One-Dimensional Spring System

The fundamental operation of the finite element method is illustrated by analysis of a one-dimensional spring system. A two-spring structure is sketched in Figure 1-2. Each spring is an element identified by the number in the box. The spring elements have a node at each end and they connect at a common node. The number in the circle labels each node. The number in the box labels each element. The subscripted u values are the node displacements, i.e., degrees-of-freedom. There is an applied force F

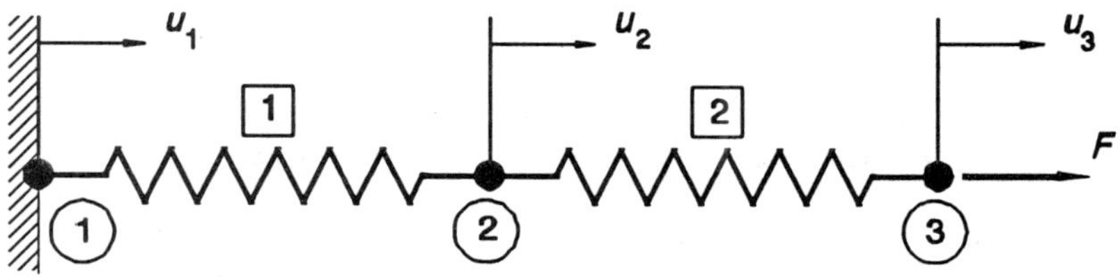

Figure 1-2. One-Dimensional Spring Structure

at node 3. We wish to solve for the node displacements and spring forces.

The first step is to formulate a general element. Figure 1-3 shows a spring element. The element label is p with nodes i and j. For a displacement formulation, assume positive displacement components of u_i at node i and u_j at node j. The element has a spring constant k, so node forces result when these displacements occur. Define f_{ip} as the force acting on node i due to the node displacements of element p. Application of simple equilibrium forms equations (1.1).

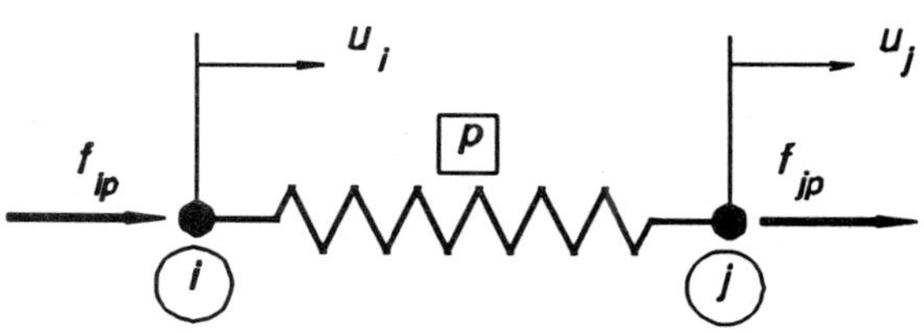

Figure 1-3. One-Dimensional Spring Element

$$\begin{aligned} f_{ip} &= -k_p u_i + k_p u_j \\ f_{jp} &= k_p u_i - k_p u_j \end{aligned} \tag{1.1}$$

We write these in matrix form in equation (1.2) and then symbolically in equation (1.3).

$$\begin{bmatrix} k_p & -k_p \\ -k_p & k_p \end{bmatrix} \begin{Bmatrix} u_i \\ u_j \end{Bmatrix} = \begin{Bmatrix} -f_{ip} \\ -f_{jp} \end{Bmatrix} \tag{1.2}$$

$$[k]\{d\} = \{f\} \tag{1.3}$$

Here, $[k]$ is the element stiffness matrix, $\{d\}$ is the element node displacement vector, and $\{f\}$ is the element node internal force vector. These steps complete the element formulation.

Now apply the general formulation to each element:

for element 1

$$\begin{bmatrix} k_1 & -k_1 \\ -k_1 & k_1 \end{bmatrix} \begin{Bmatrix} u_1 \\ u_2 \end{Bmatrix} = \begin{Bmatrix} -f_{11} \\ -f_{21} \end{Bmatrix} \tag{1.4}$$

and for element 2

$$\begin{bmatrix} k_2 & -k_2 \\ -k_2 & k_2 \end{bmatrix} \begin{Bmatrix} u_2 \\ u_3 \end{Bmatrix} = \begin{Bmatrix} -f_{22} \\ -f_{32} \end{Bmatrix} \tag{1.5}$$

The force components in the element equations are internal forces on the nodes produced by the elements when the nodes displace. Equilibrium requires that the sum of the internal forces equals the external force at each node. Representing the external force by F_i, where i represents each node, the equilibrium equations become:

$$\begin{aligned} &\text{at node 1} \quad \sum \text{forces} = 0 \quad \Rightarrow \quad -f_{11} = F_1 \\ &\text{at node 2} \quad \sum \text{forces} = 0 \quad \Rightarrow \quad -f_{21} - f_{22} = F_2 \\ &\text{at node 3} \quad \sum \text{forces} = 0 \quad \Rightarrow \quad -f_{32} = F_3 \end{aligned} \tag{1.6}$$

Substitute the element equations for the internal force terms in the equilibrium equations (1.6), and that, in effect, performs the structure assembly and yields the structure equations (1.7).

These are written in matrix form in equation (1.8) and symbolically in equation (1.9) where $[K]$ is the structure stiffness matrix, $\{D\}$ is the structure node displacement vector, and $\{F\}$ is the structure external force vector.

The set of structure or system equations must now be solved. The spring constants of the springs are known, so all terms in the structure

$$\begin{aligned} k_1u_1 - k_1u_2 \quad\quad\quad\quad\quad\quad &= F_1 \\ -k_1u_1 + k_1u_2 + k_2u_2 - k_2u_3 &= F_2 \\ - k_2u_2 + k_2u_3 &= F_3 \end{aligned} \tag{1.7}$$

$$\begin{bmatrix} k_1 & -k_1 & 0 \\ -k_1 & k_1+k_2 & -k_2 \\ 0 & -k_2 & k_2 \end{bmatrix} \begin{Bmatrix} u_1 \\ u_2 \\ u_3 \end{Bmatrix} = \begin{Bmatrix} F_1 \\ F_2 \\ F_3 \end{Bmatrix} \tag{1.8}$$

$$[K]\{D\} = \{F\} \tag{1.9}$$

stiffness matrix are known. The applied forces are known and the node displacements become the unknowns in this set of three simultaneous equations. We get the solution by premultiplying both sides of equation (1.9) by the inverse of $[K]$. However, in this case the inverse of $[K]$ is singular, meaning that we cannot get a unique solution. Physically, this means that the structure can be in equilibrium at any location in the x space, and it is free to occupy any of those positions. This allows rigid body motion. To have a unique solution we must locate the structure; that is, apply boundary conditions such as a fixed displacement on one of the nodes which is enough to prevent rigid body motion.

If an external force F applies to node 3, and the spring attaches to the wall at node 1, then it is natural to set the displacement of node 1 to zero. This action zeroes the first column of terms in the structure stiffness matrix, and that leaves three equations with two unknowns. If the value of the reaction force at node 1 is unknown, then we may skip the first equation and choose the second and third equations to solve for the unknown displacements. If the external force on node 2 is zero then

$$\begin{bmatrix} k_1+k_2 & -k_2 \\ -k_2 & k_2 \end{bmatrix} \begin{Bmatrix} u_2 \\ u_3 \end{Bmatrix} = \begin{Bmatrix} 0 \\ F \end{Bmatrix} \tag{1.10}$$

Now we may get the solution of the resulting equations (1.10) by premultiplying both sides of the equation by the inverse of this reduced structure stiffness matrix. Using the solved displacements, we calculate each element internal force by use of the individual element equations. In this example the force calculation is trivial, but in more complex elements this step determines stresses in elements of the structure.

Also, in this example the calculation of the reaction force at node 1 is easily done from elementary equilibrium equations. However, in general cases, the structure under analysis may be statically indeterminant and the reaction forces at locations of support or fixed displacement may not be known. In that situation, the equations involving these reaction forces are stored and solved after the displacements are found to calculate any desired reaction forces.

This detailed example illustrates most of the fundamental steps in the finite element method. The finite element method obviously overpowers the example case. However, a complex spring arrangement could use the procedure for analysis, or if a computer program were written, solution for a complex spring arrangement would come quickly with input of the spring constants and connectivity. The major differences between this example and actual practice are that (1) nodes usually have more than one displacement component or degree-of-freedom, (2) the element formulation is chosen to match the class of structure being analyzed, and (3) a large number of equations must be solved.

1.3 Using a Computer Program

There are three stages that describe the use of any existing finite element program. The preprocessing stage creates the model of the structure from inputs provided by the analyst. A preprocessor then assembles the data into a format suitable for execution by the processor in the next stage. The processor is the computer code that generates and solves the system equations. The third stage is postprocessing. The solution in numeric form is very difficult to evaluate except in the most simple cases. The postprocessor accepts the numeric solution, presents selected data, and produces graphic displays of the data that are easier to understand and evaluate.

Figure 1-4 draws a block diagram of a typical finite element computer program. Before entering the program's preprocessor, the user should have planned the model and gathered necessary data. In the preprocessor block, the user defines the model through the commands available in the preprocessor. The definition includes input and generation of all node point coordinates, selection of the proper element from the program's element library, input and generation of node connectivity to define all elements, input of material properties, and specifying all displacement boundary conditions, loads, and load cases. The completion of the preprocessing stage results in creation of an input data file for the analysis processor.

The processor reads from the input data file each element definition, calculates terms of the element stiffness matrix, and stores them in a data

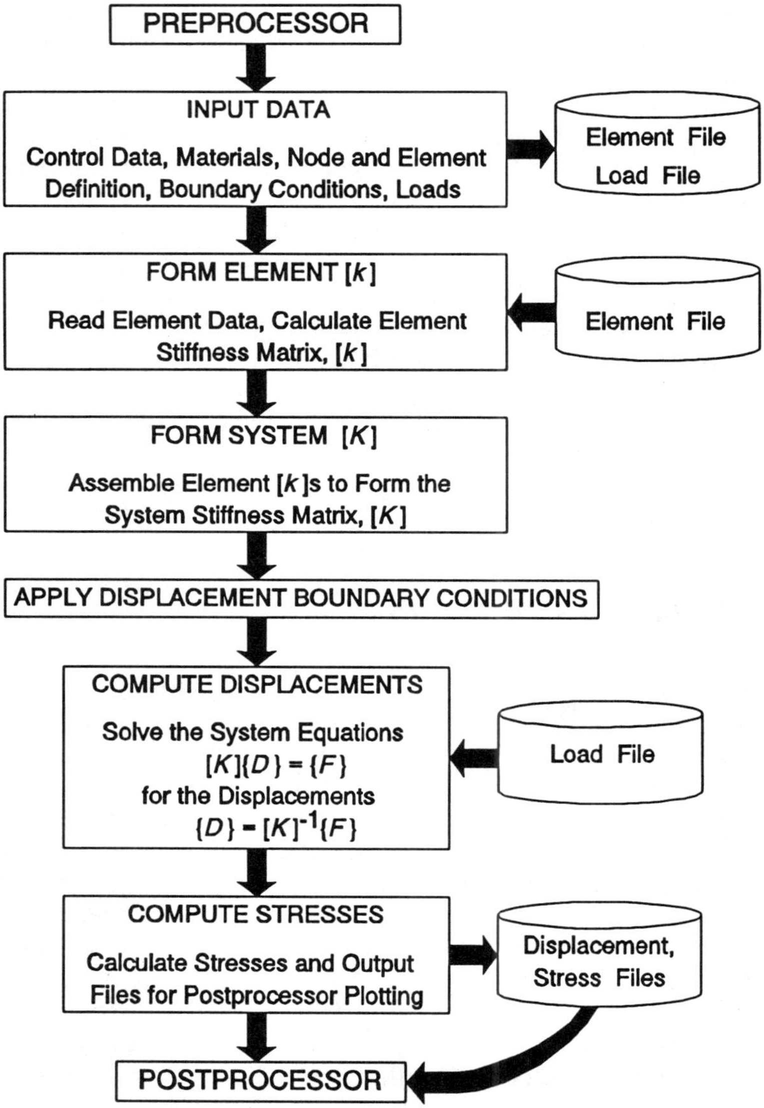

Figure 1-4. Finite Element Computer Program Block Diagram

array or on a disk file. The element type selection determines the form of the element stiffness matrix. The next step is to assemble the structure stiffness matrix by matrix addition of all element stiffness matrices. The application of enough displacement boundary conditions to prevent rigid body motion reduces the structure stiffness matrix to a nonsingular form.

Then the equation solution may be done with several different computer algorithms, but most use some variation of Gaussian elimination. For a large set of equations this is the most computationally intensive step in the process. The solution here yields values for all node point displacement components in the model. The node displacements associated with each element combined with the element formulation matrix yield the element strains. The element strains with the material properties yield the stresses in each element. The processor then produces an output listing file with data files for postprocessing.

Postprocessing takes the results files and allows the user to create graphic displays of the structural deformation and stress components. The node displacements are usually very small for most engineering structures so they are magnified to show an exaggerated shape. Node displacements are single-valued, but node values of stress are multi-valued if more than one element is attached to a given node. Node stress values are usually reached by extrapolation from internal element values and then averaged for all elements attached to the node. Contour plots or other stress plots desired by the user are created from the node values. In some post-processing programs criterion plots of the factor of safety, stress ratio to yield, or stress ratio to allowable values are also generated and displayed.

The engineer is then responsible for interpreting the results and taking whatever action is proper. The user must estimate the validity of the results first. This is very important because the tendency is to accept the results without question. Experience, thorough checking of the modeling assumptions and resulting predicted behavior, and correlation with other engineering calculations or experimental results all contribute to estimating the validity of the results.

References

1.1 Zienkiewicz, O. C. and Cheung, Y. K., *The Finite Element Method in Structural and Continuum Mechanics*, McGraw-Hill, London, 1967.

1.2 Akin, J. E., *Finite Element Analysis for Undergraduates*, Academic Press, London, 1986.

1.3 Gallagher, R. H., *Finite Element Analysis — Fundamentals*, Prentice-Hall, Englewood Cliffs, New Jersey, 1975.

1.4 Huebner, K. H. and Thornton, E. A., *The Finite Element Method for Engineers*, John Wiley and Sons, New York, 1982.

1.5 Irons, B. and Ahmad, S., *Techniques of Finite Elements*, John Wiley and Sons, New York, 1980.

1.6 Irons, B. and Shrive, N., *Finite Element Primer*, John Wiley and Sons, New York, 1983.

1.7 Reddy, J. N., *An Introduction to the Finite Element Method*, McGraw-Hill, New York, 1984.

1.8 Zienkiewicz, O. C. and Taylor, R. L., *The Finite Element Method, Volume 1 — Basic Formulation and Linear Problems*, Fourth Edition, McGraw-Hill, New York, 1989.
1.9 Cook, R. D., Malkus, D. S., and Plesha, M. E., *Concepts and Applications of Finite Element Analysis*, Third Edition, John Wiley and Sons, New York, 1989.
1.10 Fenner, D. N., *Engineering Stress Analysis: A Finite Element Approach with Fortran 77 Software*, John Wiley and Sons, New York, 1987.
1.11 Potts, J. F. and Oler, J. W., *Finite Element Applications with Microcomputers*, Prentice-Hall, Englewood Cliffs, New Jersey, 1989.
1.12 Ross, C. T. F., *Finite Element Methods in Structural Mechanics*, John Wiley and Sons, New York, 1985.
1.13 Stasa, F. L., *Applied Finite Element Analysis for Engineers*, Holt, Rinehart and Winston, New York, 1985.
1.14 Weaver, W. and Johnston, P. R., *Finite Elements for Structural Analysis*, Prentice-Hall, Englewood Cliffs, New Jersey, 1984.
1.15 Akin, J. E., *Application and Implementation of Finite Element Methods*, Academic Press, London, 1982.
1.16 Bathe, K. J. and Wilson, E. L., *Numerical Methods in Finite Element Analysis*, Prentice-Hall, Englewood Cliffs, New Jersey, 1976.
1.17 Bathe, K. J., *Finite Element Procedures in Engineering Analysis*, Prentice-Hall, Englewood Cliffs, New Jersey, 1981.

CHAPTER 2

TRUSSES

The primary focus of this text is on the aspects of finite element analysis that are more important to the user than the formulator or programmer. However, for the user to employ the method effectively he or she must have some understanding of the element formulations as well as some of the computational aspects of the programming. Therefore, the next chapters begin by looking at the element formulation for a given structural behavior class before proceeding to model development and the proper modeling approach. Most finite elements develop from use of an assumed displacement approximation; therefore, the elements presented will all deal with assumed displacement formulations.

2.1 Direct Element Formulation

This section presents the direct physical formulation of a truss element and its spatial orientation to solve two-dimensional frameworks. A member of a truss structure is like a one-dimensional spring. The member has a length substantially larger than its transverse dimensions, and it has a pinned connection to other members that eliminates all loads other than axial load along the member length. It usually has a constant cross-section area and modulus of elasticity along its length. The stiffness is then

$$k = \frac{AE}{L} \tag{2.1}$$

where A is the cross-section area, E is the modulus of elasticity, and L is the member length.

A one-dimensional truss element then has an element formulation identical to the one-dimensional spring given in equation (1.2). This is its element stiffness matrix for one-dimensional displacement and loading along the axis of the member. For member positioning in a two-dimensional space as illustrated in Figure 2-1, each node has two components of displacement u and v and two components of force p and q. This leads to a set of element equations with an element stiffness matrix of size 4 by 4.

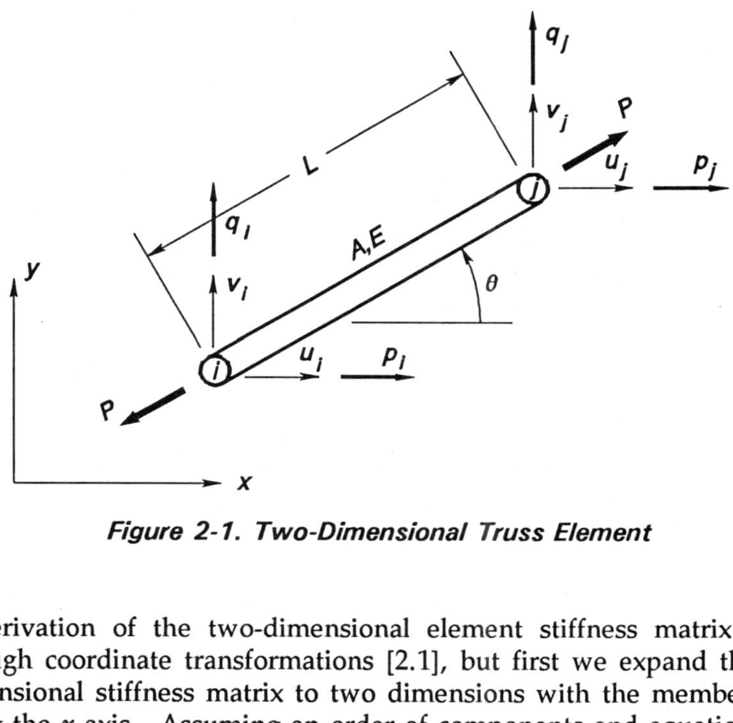

Figure 2-1. Two-Dimensional Truss Element

Derivation of the two-dimensional element stiffness matrix comes through coordinate transformations [2.1], but first we expand the one-dimensional stiffness matrix to two dimensions with the member lying along the x axis. Assuming an order of components and equations of u and v at node i followed by u and v at node j, the element equations are written in equation (2.2).

$$\begin{bmatrix} k & 0 & -k & 0 \\ 0 & 0 & 0 & 0 \\ -k & 0 & k & 0 \\ 0 & 0 & 0 & 0 \end{bmatrix} \begin{Bmatrix} u_i \\ v_i \\ u_j \\ v_j \end{Bmatrix} = \begin{Bmatrix} -p_i \\ -q_i \\ -p_j \\ -q_j \end{Bmatrix} \tag{2.2}$$

Notice that the terms relating displacement and force in the x direction are the spring constant of the member and the terms relating displacement and force in the y direction are zero. A linear analysis always assumes that the displacements are much smaller than the overall geometry of the structure; therefore, the stiffness is based on the undeformed configuration. In this case, because it is a motion perpendicular to the line of the member, if we consider a vertical displacement component at one of the nodes, no vertical force results because there is no axial stretch relative to the undeformed configuration.

This formulation represents the element stiffness matrix in a local element coordinate system that is aligned with the element axis. To position the element at an arbitrary angle, θ, from the x coordinate axis, we perform a transformation of coordinate systems to derive the element stiffness matrix in the x,y global coordinate system. In the system of equations, the displacements and forces are both vectors, so they transform through standard vector transformations. The displacement components in global coordinates relate to local components through equation (2.3).

$$\{d\} = [T]\{d'\} \tag{2.3}$$

Here, $\{d'\}$ are the global displacement components, $[T]$ is the transformation matrix, and $\{d\}$ are the local element coordinate displacement components.

The transformation matrix is given by equation (2.4).

$$[T] = \begin{bmatrix} c & s & 0 & 0 \\ -s & c & 0 & 0 \\ 0 & 0 & c & s \\ 0 & 0 & -s & c \end{bmatrix} \tag{2.4}$$

Here, s is the $\sin\theta$, and c is the $\cos\theta$. Similarly, the force components in the global coordinate system are given by

$$\{f\} = [T]\{f'\} \ . \tag{2.5}$$

The element stiffness matrix in the local coordinate system is defined in matrix notation from equation (2.2) by

$$[k]\{d\} = \{f\} \ . \tag{2.6}$$

Making the substitutions for $\{d\}$ and $\{f\}$ given above yields

$$[k][T]\{d'\} = [T]\{f'\} . \tag{2.7}$$

The transformation matrix is an orthogonal matrix, meaning that

$$[T]^{-1} = [T]^{T} . \tag{2.8}$$

Therefore, multiplying equation (2.7) by $[T]^{T}$ produces

$$[T]^{T}[k][T]\{d'\} = \{f'\} \tag{2.9}$$

which makes

$$[k'] = [T]^{T}[k][T] = k \begin{bmatrix} c^2 & cs & -c^2 & -cs \\ cs & s^2 & -cs & -s^2 \\ -c^2 & -cs & c^2 & cs \\ -cs & -s^2 & cs & s^2 \end{bmatrix} \tag{2.10}$$

The use of this element formulation and equation assembly is shown through the example truss structure pictured in Figure 2-2. The elements and nodes are numbered, and load and boundary conditions are shown. The structure equations are

$$[K]\{D\} = \{F\} \tag{2.11}$$

where, $[K]$ is the structure stiffness matrix, $\{D\}$ is the node displacement vector, and $\{F\}$ is the applied load vector.

These equations come from applying the conditions of equilibrium to all the nodes by setting the summation of internal forces equal to the applied forces. The internal forces are given by the product of each element stiffness matrix with its node displacements. This yields equation (2.12),

$$[k]\{d\}|_1 + [k]\{d\}|_2 + [k]\{d\}|_3 = \{F\} \tag{2.12}$$

where the subscripts refer to the numbered elements. If the displacement vector in each term of the equation above was identical, then we could

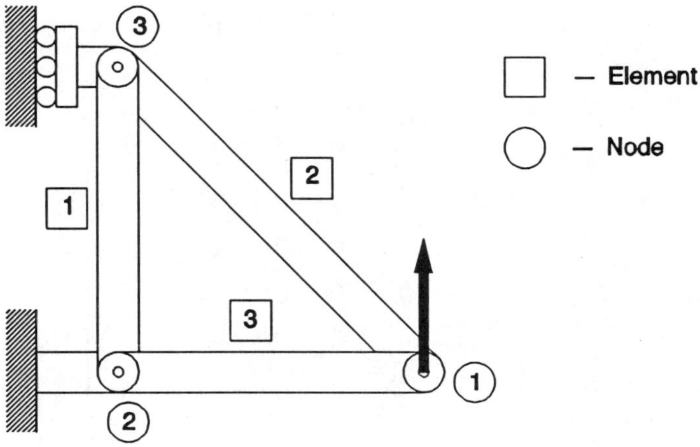

Figure 2-2. Example Truss Structure

factor it out and add the stiffness matrices term-by-term to produce the structure stiffness matrix.

The displacement vector for each element must then expand to include all the structure degrees-of-freedom, not just the ones associated with a given element. In order for the matrix equation to be correct, a corresponding expansion of the element stiffness matrix must accompany the expansion of the displacement vector. It expands to the size of the structure stiffness matrix which in this example becomes a 6-by-6 matrix. The expansion simply adds rows and columns of zeroes to each element stiffness matrix corresponding to the additional structure degrees-of-freedom unused in the given element [2.1].

Applying this approach, the stiffness matrix for element 1 in the example results from using equation (2.10) with a θ value of 90 degrees. Rows and columns of zeroes fill in equations and positions involving u_1 and v_1 as shown in equation (2.13).

$$[k]_1 = \begin{bmatrix} 0 & 0 & 0 & 0 \\ 0 & k_1 & 0 & -k_1 \\ 0 & 0 & 0 & 0 \\ 0 & -k_1 & 0 & k_1 \end{bmatrix} = \begin{bmatrix} 0 & 0 & 0 & 0 & 0 & 0 \\ 0 & 0 & 0 & 0 & 0 & 0 \\ 0 & 0 & 0 & 0 & 0 & 0 \\ 0 & 0 & 0 & k_1 & 0 & -k_1 \\ 0 & 0 & 0 & 0 & 0 & 0 \\ 0 & 0 & 0 & -k_1 & 0 & k_1 \end{bmatrix} \quad (2.13)$$

Similarly, the matrix for element 2 with θ equal to 135 degrees and rows and columns filled in equations and positions involving u_2 and v_2 results in equation (2.14).

$$[k]_2 = \begin{bmatrix} .5k_2 & -.5k_2 & 0 & 0 & -.5k_2 & .5k_2 \\ -.5k_2 & .5k_2 & 0 & 0 & .5k_2 & -.5k_2 \\ 0 & 0 & 0 & 0 & 0 & 0 \\ 0 & 0 & 0 & 0 & 0 & 0 \\ -.5k_2 & .5k_2 & 0 & 0 & .5k_2 & -.5k_2 \\ .5k_2 & -.5k_2 & 0 & 0 & -.5k_2 & .5k_2 \end{bmatrix} \tag{2.14}$$

Finally for element 3, θ is 0 degrees and u_3 and v_3 are the additional degrees-of-freedom in equation (2.15).

$$[k]_3 = \begin{bmatrix} k_3 & 0 & -k_3 & 0 & 0 & 0 \\ 0 & 0 & 0 & 0 & 0 & 0 \\ -k_3 & 0 & k_3 & 0 & 0 & 0 \\ 0 & 0 & 0 & 0 & 0 & 0 \\ 0 & 0 & 0 & 0 & 0 & 0 \\ 0 & 0 & 0 & 0 & 0 & 0 \end{bmatrix} \tag{2.15}$$

The summations of equation (2.12) are now carried out by adding the expanded element stiffness matrices term-by-term. The resulting structure stiffness matrix is in equation (2.16).

$$[K] = \begin{bmatrix} .5k_2+k_3 & -.5k_2 & -k_3 & 0 & -.5k_2 & .5k_2 \\ -.5k_2 & .5k_2 & 0 & 0 & .5k_2 & -.5k_2 \\ -k_3 & 0 & k_3 & 0 & 0 & 0 \\ 0 & 0 & 0 & k_1 & 0 & -k_1 \\ -.5k_2 & .5k_2 & 0 & 0 & .5k_2 & -.5k_2 \\ .5k_2 & -.5k_2 & 0 & -k_1 & -.5k_2 & .5k_2+k_1 \end{bmatrix} \tag{2.16}$$

Before solving the equations, apply the displacement boundary conditions. In this example the boundary conditions have components u_2, v_2, and u_3 equal to zero. Applying these to the system equations zeroes the third, fourth, and fifth columns of the structure stiffness matrix. This leaves six equations with three unknown displacements. In most problems, the reaction forces in equations 3, 4, and 5 are also unknown, so choose equations 1, 2, and 6 to solve for the displacement components, u_1, v_1, and v_3.

We get the solution by finding the inverse of the remaining 3-by-3 stiffness matrix. Multiplying the inverse with the load vector yields the displacements. After finding the displacements, calculate the element forces by use of the element equations.

This concludes the example and demonstration of two-dimensional truss formulation. Of course, this extends easily to three dimensions. The assembly of equations by expanding the element stiffness matrix to structure size is useful for explanation of the process but impractical for a large number of system equations. In computer programs the algorithm only needs to place the terms of the element stiffness matrix in the correct position in the structure stiffness matrix. This is easy to accomplish because the structure equation sequence correlates to the node number. Elements are defined by node numbers, thus providing the direct correlation for positioning the terms.

2.2 *The Finite Element Model*

The practitioner of finite element analysis normally uses an existing computer code. Because of the general complexity and sizable effort required to create a finite element code, it is impractical to consider writing a code for every specific problem that needs solving. So the job of the practitioner is to use an existing code to solve the problem of interest. The user in this case is responsible for creating the model of the structure, for managing the execution of the program, and for interpreting results created by the program.

In following the analysis procedure we reach the point of planning the model. The arrangement of nodes and elements that describe the model is known as the mesh. Using all the information that is known about the problem, and knowing the capabilities of the program chosen to analyze the problem, the mesh is planned to model the structure properly. In truss structures, each member is modeled as one truss element with the connections of truss members or elements at the node points. Based on the formulation and general assumptions for truss members, these node connections behave as pinned joints. This results in no flexural loading of the member in that one member can swivel or hinge relative to the others connected at any given node, yet will transmit axial load.

Since a truss element behaves exactly in agreement with the assumptions of a truss member, there is no need to divide a member into more than one element. In fact, such a subdivision will cause the execution stage of the program to fail. The failure is due to the zero stiffness against any lateral force applied at a node connection where two members are in perfect axial alignment. So just as a physical truss structure constructed in this manner would collapse, the numerical solution of the problem defined in this manner also should collapse.

A simple bridge structure is in Figure 2-3 and a corresponding finite element mesh with elements and nodes numbered is in Figure 2-4.

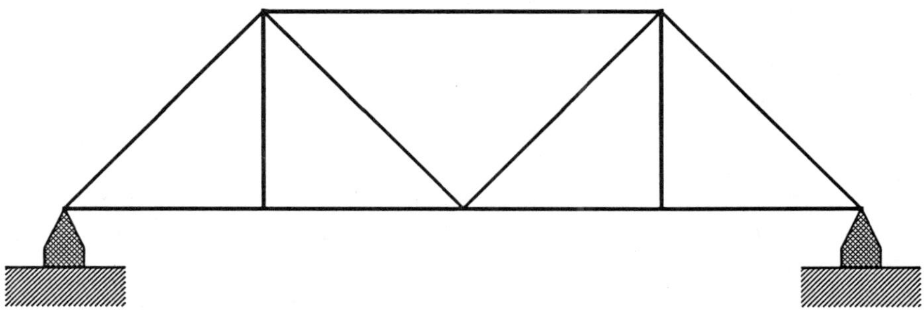

Figure 2-3. Simple Bridge Structure

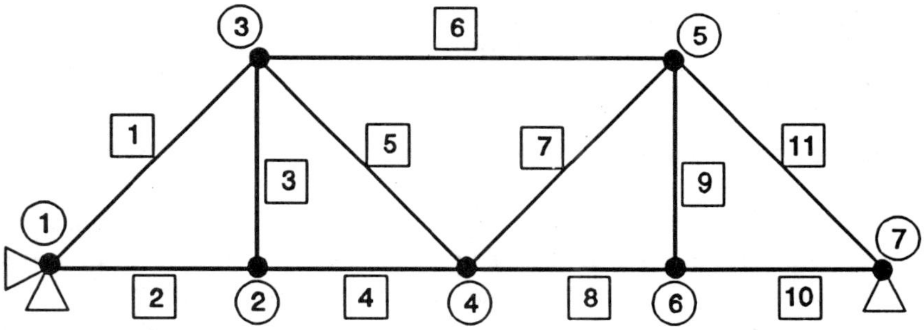

Figure 2-4. Simple Bridge Structure Finite Element Mesh

2.3 The Analysis Step

Operation of the analysis step is mostly transparent to the user. However, there are some factors that the user must be familiar with to assure good results. Some of them are inherent in the computer hardware and software, but some of them are under user control. For small truss structure models, most programs have enough numerical accuracy and performance to provide an accurate solution without much user concern. Large models cause the most concern [2.1].

A large model is one in which there are many elements and nodes used to represent the structure. (The structure itself is not necessarily large.) With many system equations it becomes difficult to find a numerical solution if the equation matrix is full. Even the inverse of a 10-by-10 matrix may be inaccurate if done by Gauss elimination with only a few significant figures carried along in the mathematical operations. The accuracy will improve, however, if the matrix has its nonzero terms clustered near the diagonal. This reduces the number of operations and reduces the roundoff error carried along in each operation. This kind of matrix, with its nonzero terms near the diagonal, is a banded matrix.

If the truss structure model consists of thousands of nodes and elements, then the bandwidth of the structure equations needs to be small. Keeping it small reduces error and computing time. If the mesh plan does not have a small bandwidth for the system of equations, then bandwidth minimizers available in many programs should reduce it. There are several algorithms available which will usually, but not always, find a better node or element numbering pattern. Most programs keep the original numbering in the model for documentation and presentation purposes by storing the renumbered nodes and elements in translation tables. In some programs the user has the option to keep the original numbering or change to the new numbering plan.

The approximation error for the truss element is zero since the element formulation is in exact agreement with the assumptions used to define a truss member. During processor execution there are usually some prompts of progress made displayed on the computer, and if errors occur, messages appear. Sometimes these messages have meaning only to the computer program developer, but some of them can be very helpful in determining the model error.

The most common runtime errors involve incorrect definition of elements or incorrect application of displacement boundary conditions. For example, both conditions can produce an error message that the structure stiffness matrix is not positive-definite or that a negative pivot or diagonal term in the stiffness matrix appeared during equation reduction. For truss models this can occur whenever there are not enough boundary conditions to prevent rigid body motion. It can also occur when two elements connect in-line resulting in zero lateral stiffness. It can also mean that the truss structure itself is not kinematically stable associated with a kinematic linkage of the members.

The error messages from the computer program should provide an associated element number, node number, or equation number where the error occurred. That makes it easier for the user to pinpoint the problem. When execution completes without errors, then postprocessing may begin.

2.4 Output Processing and Evaluation

At this stage of the analysis, all programs have numerical results in the form of a listing file of the problem. This file will include a summary of the input data followed by numerical values of all node displacement components and all element stress results. One of the important steps to take here is to review the summary of the input printout, scanning it for errors in input interpretation of the data entered or selection of default parameters that are not appropriate for the problem at hand. This can usually be done effectively if the program formats the data for easy viewing.

The results for displacements and stresses in the listing file for large models are so lengthy that scanning is not practical. However, many programs will print a summary of maximum values for displacement components and stress magnitudes. Therefore, it is very desirable to present the data graphically for more effective evaluation.

The first graphic of importance should be an exaggerated deformed shape of the structure. All postprocessing programs will include this graphic that uses the node displacements with a scale factor to exaggerate the deformation and make it more apparent to the eye. The deflections in most engineering structures are usually very small, and without an exaggeration scale factor the deformed shape would look the same as the undeformed shape. Program options usually exist either to provide both an undeformed and deformed mesh simultaneously or an outline of the undeformed object superimposed on the graphic of the deformed mesh.

The engineer must look at this plot critically and make sure that the boundary conditions are correct and that the shape of the deformed structure agrees qualitatively with the expected deformation. In truss structures the deformed shape will obviously show each member or element as a straight line connecting the nodes in an exaggerated deformed position.

After thorough evaluation of the deformed shape, the graphics should then turn to plots of the stress components. In continuum structures the stress component plots relate to averaged quantities at the node points. Truss structures have a stress in each member that is constant, and most commercial postprocessing programs do not provide much in the way of graphic presentation of these stresses. In this event the user must return to the listing file and examine it for the highest stressed members.

The evaluation of the results determines whether we need to make additional model refinements and whether the results have converged to enough accuracy. For truss structures we know that the element formulation is exact; therefore, in a correctly defined and processed model the output results will be exact. So there is no need for refined modeling to produce converged results in this case, but the modeling of loads and boundary conditions may not be fully appropriate.

It is important to remember that this is a linear elastic analysis. One of the potential failure modes is overstressing while another is elastic buckling. The stresses compare to yield strength for the material to determine if overstress failure occurs. To determine whether there is potential for elastic buckling, the user must identify the members with a significant compression load, and then use Euler buckling equations from mechanics of materials to evaluate the potential for each member to buckle [2.2]. This is not done in a linear elastic analysis computer program. If any member has an inadequate safety factor against buckling, then the entire structure should have a stability analysis conducted using a solution algorithm available in some nonlinear computer codes.

2.5 Case Study

We show a typical truss structure in Figure 2-5. This structure has members of two different cross-section areas and two different materials. The mesh plan for this structure is illustrated in Figure 2-6 with a chosen

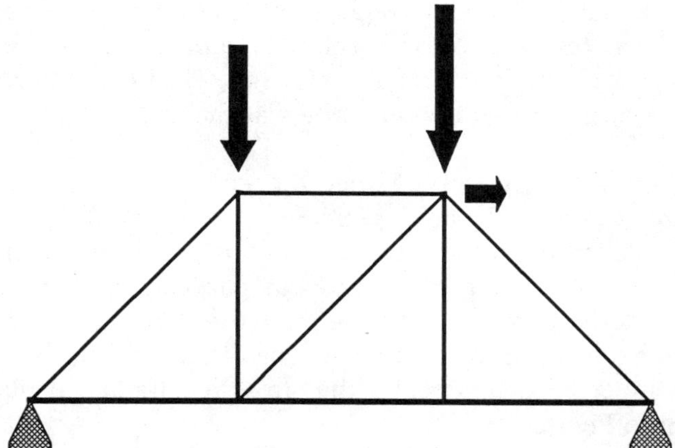

Figure 2-5. Two-Dimensional Truss Structure

pattern for node and element numbers. The enforced boundary conditions symbolized by the triangular shapes have the triangle tip placed on the node pointing in the direction of restraint. The loads symbolized by long arrows apply to the indicated node points.

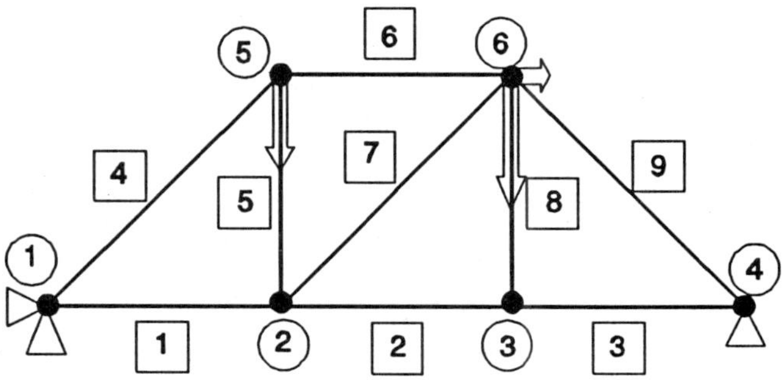

Figure 2-6. Truss Structure Finite Element Model

Table 2-1 lists the input data required to create this model. The title line is first followed by the control data line. Node definition lines begin with their number, with boundary condition restraints and coordinate locations following. In this model all the z boundary conditions are restrained and all the z coordinates are zero because this is a 2-D truss element without any z degrees-of-freedom. Additional control data is next followed by load data lines. The load data lines begin with the node number of application with a direction and magnitude. The type of element is a truss. The material data is shown with two table entries. Material 1 has a modulus of elasticity for steel with a cross-section area of 0.4 in^2, and material 2 is aluminum with a cross-section area of 0.7 in^2. The element definitions are given by the entry of two node numbers at the endpoints of the element with a material table assignment.

This is such a small model that the equation bandwidth factor is insignificant along with any question of numerical performance or precision. Following program execution the deformed shape of the structure is shown in Figure 2-7. Note particularly that the enforced boundary conditions match, and the structure deforms in a manner that agrees with its expected deformation. The displacement, load, and stress results are in Table 2-2. Finally, the stress results are displayed in bar graph form in Figure 2-8.

We may evaluate these results for the design by knowing the additional property given by its yield strength. If we say the steel has a yield strength of 50 kpsi and the aluminum has a yield strength of 30 kpsi, then the lowest factor of safety will be in element number 9 which has a value

Table 2-1. Input Data File for Truss Structure Model

```
************|notes inside vertical bars are for data explanation only|
Truss Case Study                                          |title line|
  6   1    1             |6 nodes, 1 element group, 1 load case|
  1   1  1  1    0.000   0.000    0.000          |node number,    |
  2   0  0  1   10.000   0.000    0.000          |x,y,z boundary|
  3   0  0  1   20.000   0.000    0.000          |conditions,     |
  4   0  1  1   30.000   0.000    0.000          |0-free,1-fixed|
  5   0  0  1   10.000  10.000    0.000          |x,y,z           |
  6   0  0  1   20.000  10.000    0.000          |coordinates     |
  0                      |number of inclined boundary conditions|
  1   3                                     |load case 1, 3 loads|
  5   2  -5000.0                      |node 5, y dir., -5000 value|
  6   1   2000.0                      |node 6, x dir.,  2000 value|
  6   2  -7000.0                      |node 6, y dir., -7000 value|
  1   9    2         |element type 1-truss, 9 elements, 2 materials|
  1   30.0E+06  0.40                |material 1, E = 30E6, A = 0.4|
  2   10.0E+06  0.70                |material 2, E = 10E6, A = 0.7|
  1   1    2    1        |element #, node I, node J, material #|
  2   2    3    1
  3   3    4    1
  4   1    5    2
  5   5    2    2
  6   5    6    2
  7   6    2    2
  8   3    6    2
  9   6    4    2
```

Figure 2-7. Truss Structure Deformed Shape

of 2.12. We see that members 4, 6, and 9 have compressive axial loads. If we use the Euler buckling equation we have to know more than just its elastic modulus and cross-section area. We also must know its cross-section shape to find the area moment of inertia and the end conditions. If we assume pinned end conditions and the cross-section shape is a solid circular rod then the buckling loads for these elements would be 19250, 38500, and 19250 respectively. Thus, we also have a reasonable factor of safety against buckling.

Table 2-2. Results Data File for Truss Structure Model

D I S P L A C E M E N T S

NODE	X-DISP	Y-DISP	Z-DISP
1	0.000000	0.000000	0.000000
2	0.005833	-0.035436	0.000000
3	0.011667	-0.037694	0.000000
4	0.017500	0.000000	0.000000
5	0.015233	-0.035436	0.000000
6	0.008091	-0.037694	0.000000

S T R E S S E S I N T R U S S E L E M E N T
G R O U P 1

ELEM #	FORCE	STRESS
1	7000.	17500.
2	7000.	17500.
3	7000.	17500.
4	-7071.	-10102.
5	0.	0.
6	-5000.	-7143.
7	0.	0.
8	0.	0.
9	-9899.	-14142.

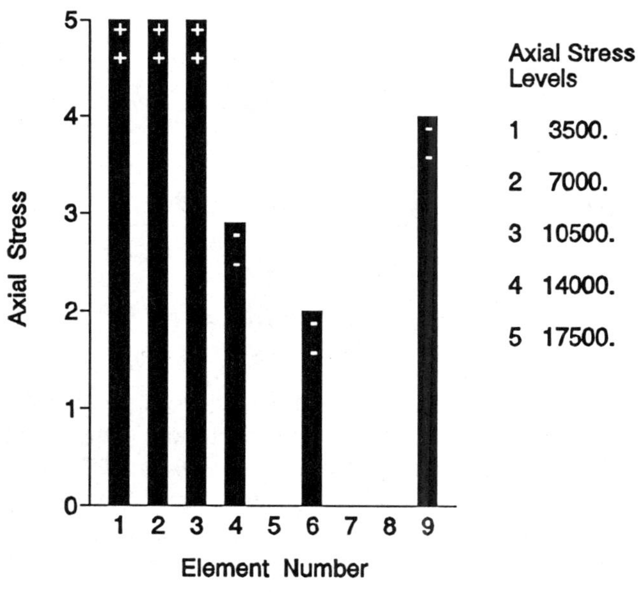

Figure 2-8. Truss Structure Member Stresses

2.6 *Closure*

There are few situations in mechanical design where a truss element is the right element for modeling the behavior. It is a simple element with which to discuss and learn finite element concepts. It may make an important contribution to the analysis by use as a boundary supporting spring or a gap element that connects two or more separate parts of a machine that must interact in the analysis. Therefore, the engineer must understand its nature well to interpret its effect on the overall response of any analysis that includes truss or truss-based elements.

Problems

2.1 A three-member truss structure is shown in Figure P2-1 with corresponding node and element numbering for a finite element model. Elements 1 and 2 are aluminum, and element 3 is steel. The cross-section areas are 1.5 sq. in. for element 1 and 1.0 sq. in. for elements 2 and 3. Determine the displacement of node 2 and the stresses in each member. Solve by use of a computer program and by hand calculation. Report in a neat and concise informal engineering communication. Please include the hand-calculated structure stiffness matrix in its full (8-by-8) and reduced form.

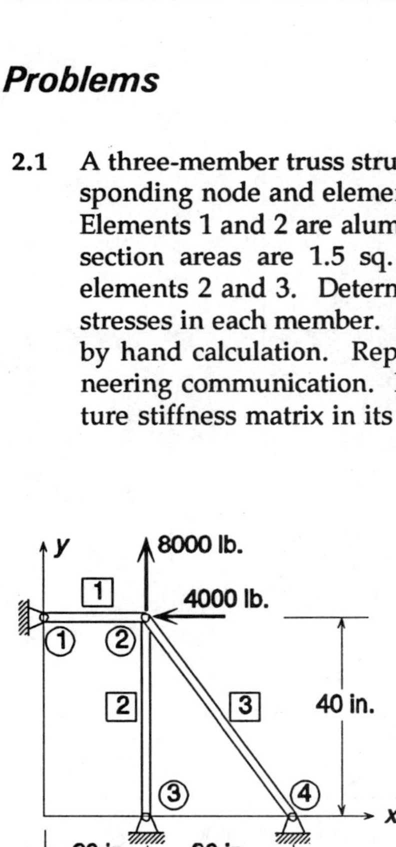

Figure P2-1.

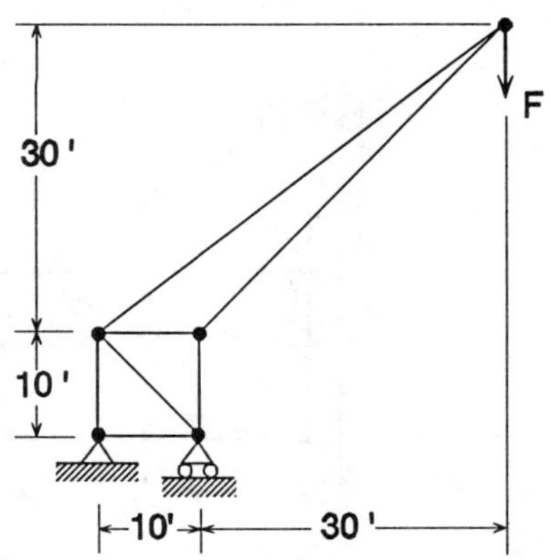

Figure P2-2.

2.2 Design the derrick structure shown for a load capacity of 20 kips. Choose a suitable steel and, using a factor of safety of 4.0, determine the cross-section area for all the members. Recommend a cross-section shape that will prevent any member from buckling.

2.3 Design a cantilevered boom to support the loads shown in Figure P2-3. All members are steel with a cross-section area of 1 sq. in. The material has an allowable stress of 20 kpsi. First determine if the design is satisfactory as illustrated. Next redesign the structure within the geometric boundaries shown and the same allowable stress. A redesign may change the member arrangements, eliminate members, or change cross-section areas. One of the redesign goals should be to reduce the overall weight of the structure. Determine a suitable cross-section shape to prevent buckling for each member in compression.

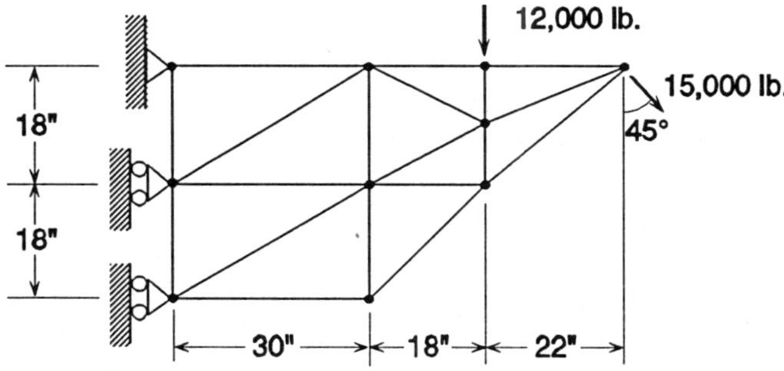

Figure P2-3.

2.4 Repeat Problem 2.3 for the boom in Figure P2-4.

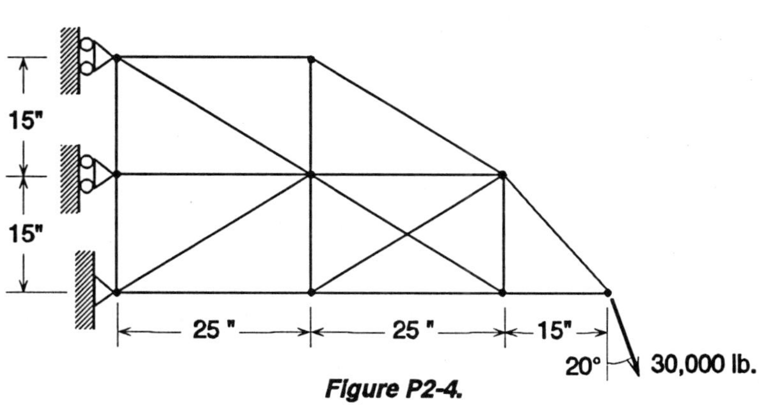

Figure P2-4.

References

2.1 Cook, R. D., Malkus, D. S., and Plesha, M. E., *Concepts and Applications of Finite Element Analysis*, Third Edition, John Wiley and Sons, New York, 1989.

2.2 Popov, E. P., *Introduction to Mechanics of Solids*, Prentice-Hall, Inc., Englewood Cliffs, New Jersey, 1968.

CHAPTER 3

BEAMS AND FRAMES

Application of straight beam theory readily solves simple beam problems especially if the problem is statically determinant. If the beam is not particularly simple, in that it may have cross-section changes, multiple supports, or complex loading distributions, then we can use beam theory, but it is very tedious to develop the solution by hand. Also, many 2-D or 3-D framework structures may require solutions in which the truss member assumption is inadequate and therefore needs the beam flexure formulation. Further applications may include beam members as re-inforcement members in combination beam, plate, and shell structures. These applications are readily attacked with the finite element formulation.

3.1 Element Formulation

Here we follow the direct approach for formulating the element stiffness matrix [3.1]. The element equations relating general displacement and force components are given by

$$[k]\{d\} = \{f\} \tag{3.1}$$

where $[k]$ is the element stiffness matrix, $\{d\}$ is the node displacement component column matrix, and $\{f\}$ is the internal force component column matrix. The stiffness matrix terms derive from superposition of simple beam solutions. Apply a unit displacement of one component with the other components held to zero and evaluate the magnitude of resulting force components. For example, taking the element shown in Figure 3-1 and applying a unit vertical displacement v_i with $\phi_i = v_j = \phi_j = 0$ results in the deformation illustrated in Figure 3-2.

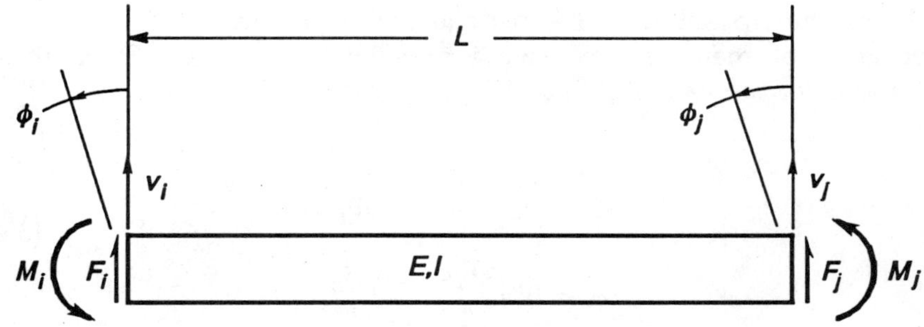

Figure 3-1. Beam Element

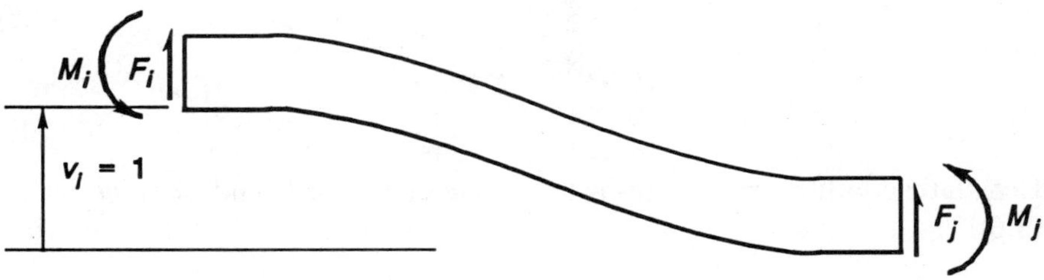

Figure 3-2. Deformed Beam Element

The forces shown, which are the superposition of the solutions for a cantilever beam with an end load and an end moment, produce this deformation. The superposition is done to yield a unit value of lateral displacement with a zero slope at the end. The element equations written in matrix form yield equation (3.2), which in turn yields the relations in equation (3.3).

$$\begin{bmatrix} k_{11} & k_{12} & k_{13} & k_{14} \\ k_{21} & k_{22} & k_{23} & k_{24} \\ k_{31} & k_{32} & k_{33} & k_{34} \\ k_{41} & k_{42} & k_{43} & k_{44} \end{bmatrix} \begin{Bmatrix} 1 \\ 0 \\ 0 \\ 0 \end{Bmatrix} = \begin{Bmatrix} F_i \\ M_i \\ F_j \\ M_j \end{Bmatrix} \tag{3.2}$$

$$k_{11} = F_i\,,\quad k_{21} = M_i\,,\quad k_{31} = F_j\,,\ \text{and}\quad k_{41} = M_j \tag{3.3}$$

Using superposition of beam deflection equations available in any mechanics of materials text, we write equations (3.4). Solve these equations for the values of F_i and M_i in equations (3.5).

$$\begin{aligned} v_i &= 1 = \frac{F_i L^3}{3EI} - \frac{M_i L^2}{2EI} \\ \phi_i &= 0 = \frac{F_i L^2}{2EI} - \frac{M_i L}{EI} \end{aligned} \tag{3.4}$$

$$\begin{aligned} F_i &= \frac{12EI}{L^3} \\ M_i &= \frac{6EI}{L^2} \end{aligned} \tag{3.5}$$

Use static equilibrium equations to get the values of F_j and M_j in equation (3.6).

$$F_j = -\frac{12EI}{L^3} \qquad M_j = \frac{6EI}{L^2} \tag{3.6}$$

We now have all the terms of column 1 of the 4x4 element stiffness matrix as shown in equation (3.7).

$$[k] = \begin{bmatrix} \frac{12EI}{L^3} & \cdots & \cdots & \cdots \\ \frac{6EI}{L^2} & \cdots & \cdots & \cdots \\ -\frac{12EI}{L^3} & \cdots & \cdots & \cdots \\ \frac{6EI}{L^2} & \cdots & \cdots & \cdots \end{bmatrix} \tag{3.7}$$

Similarly applying a unit value of rotation for ϕ_i and fixing all other components to zero, we derive the force and moment values in Figure 3-3. These come from superposition of the same solutions for end load and moment to satisfy the displacement conditions.

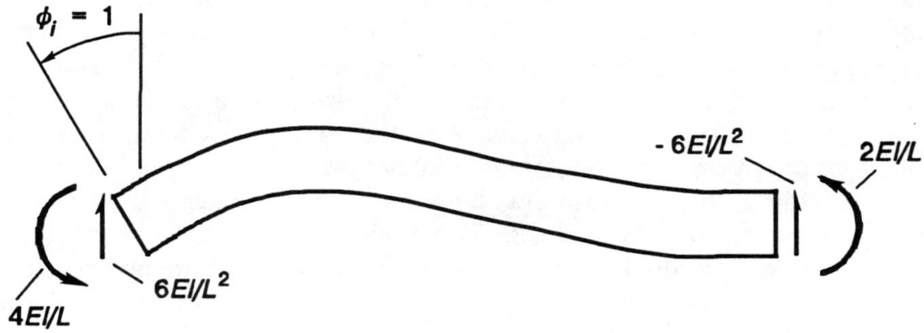

Figure 3-3. Deformed Beam Element

Notice that the sign convention employed here is common in the finite element formulation such that the component's sign always agrees with the positive direction of a right-handed coordinate system. This does not agree with most beam sign conventions employed in mechanics of material texts. Therefore, the user should be aware that the output components will normally be expressed using this finite element sign convention. This means, for example, that a positive value of moment at the first node of the element will produce a tensile stress at the top surface of the beam. In contrast, a positive moment on the second node will produce a compressive stress at the top surface of the beam.

Obtain the remaining terms in the stiffness matrix by application of the same procedures to the second node. The final element stiffness is then given in equation (3.8).

$$[k] = \begin{bmatrix} \frac{12EI}{L^3} & \frac{6EI}{L^2} & -\frac{12EI}{L^3} & \frac{6EI}{L^2} \\ \frac{6EI}{L^2} & \frac{4EI}{L} & -\frac{6EI}{L^2} & \frac{2EI}{L} \\ -\frac{12EI}{L^3} & -\frac{6EI}{L^2} & \frac{12EI}{L^3} & -\frac{6EI}{L^2} \\ \frac{6EI}{L^2} & \frac{2EI}{L} & -\frac{6EI}{L^2} & \frac{4EI}{L} \end{bmatrix} \tag{3.8}$$

This formulation provides an exact representation of a beam span within the assumptions involved in straight beam theory, provided there are no loads applied along the span. Therefore, in modeling considerations, place a node at all locations where concentrated forces, or moments, act in creating the element assembly. In spans where there is a distributed load, the assumed displacement field does not completely satisfy the governing differential equation; therefore, the solution is not exact but approximate. One approach to modeling in this area is to make enough subdivisions of the span with distributed load to lessen the error. If a work equivalent load set acting on the nodes replaces the distributed load, then the influence of any error in this element will not propagate to other elements. In other words, the displacement components at the nodes will be correct if we use the equivalent load set. The equivalent load components for a distributed load on the element span are the negative of the end reaction force and moment found in the solution of a fixed end beam with the same distributed load as shown by Logan [3.2].

This formulation provides the ability to analyze simple beams, but does not account for the axial load that may exist in beam members connected in a framework. By adding the truss element formulation by superposition with the previous formulation, we have an element that can support both lateral and axial loads. The axial stiffness terms at each node are added to the element stiffness matrix formulation to create the frame element stiffness matrix in equation (3.9).

$$[k] = \begin{bmatrix} \frac{AE}{L} & 0 & 0 & -\frac{AE}{L} & 0 & 0 \\ 0 & \frac{12EI}{L^3} & \frac{6EI}{L^2} & 0 & -\frac{12EI}{L^3} & \frac{6EI}{L^2} \\ 0 & \frac{6EI}{L^2} & \frac{4EI}{L} & 0 & -\frac{6EI}{L^2} & \frac{2EI}{L} \\ -\frac{AE}{L} & 0 & 0 & \frac{AE}{L} & 0 & 0 \\ 0 & -\frac{12EI}{L^3} & -\frac{6EI}{L^2} & 0 & \frac{12EI}{L^3} & -\frac{6EI}{L^2} \\ 0 & \frac{6EI}{L^2} & \frac{2EI}{L} & 0 & -\frac{6EI}{L^2} & \frac{4EI}{L} \end{bmatrix} \tag{3.9}$$

This assumes that superposition is valid for this case. If displacements are small it will be accurate; however, there is an interaction that occurs between axial and lateral loading on beams. If the axial load is tensile it reduces the effect of lateral loads, and when the axial load is compressive it amplifies the effect of lateral loads. To gain further information on this interaction, consult an advanced mechanics of materials text [3.3] for the equations that apply to members called beam-columns or struts. The equations for these members are a nonlinear function of the size of lateral displacement. Therefore, a linear analysis cannot account for the effect.

The user should be aware of this consideration. Remember that if the axial load is tensile, the results from beam elements will be higher than they actually are; thus results are conservative. Also, if the axial load is compressive, the results will be less than actual and may be in serious error. The size of error associated with the compressive loading is normally quite small until the axial load exceeds roughly 25 percent of the Euler column buckling load. In most cases a design should have a factor of safety against buckling greater than four anyway.

Now the formulation includes the u and v displacement components and the section rotation at the nodes in the element local coordinate system. Using the coordinate transformations developed for truss members, we may orient this two-dimensional beam element in 2-D space. Through this transformation, then, the element formulation applies to any 2-D framework.

3.2 The Finite Element Model

In planning the mesh for a structure to be modeled with beam elements, the factors just revealed in element formulation provide guidance about the proper element subdivision and connections. Since the element formulation is exact for a beam span with no intermediate loads, then we need only one element to model any member of the structure that has constant cross-section properties and no intermediate loads. Where a span has a distributed load, we may subdivide it with several elements to lessen the error depending on the solution accuracy desired.

There should be a node placed at every location in the structure where a point load is applied. Also, where frame members connect such that the line element changes direction or cross-section properties change, we should place a node and end an element at that point. Remember that the connection of two or more elements at a node guarantees that each element connecting at that node will have the same value of linear and rotation displacement components at that node. Physically, think of this as a solid, continuous, or welded configuration.

3.3 Output Processing and Evaluation

A complete printout, or listing file, lists a reflection of model input data, the displacement results including rotations, and output of stresses resulting from moment, axial, and shear forces. The graphical presentation of the deformed shape ideally would use the rotations at the nodes with the assumed displacement shape function for the element to plot the actual curved shape the elements take when loaded. However, most programs only plot the deformed shape using the node translation displacements and straight line connections to represent the elements. In this case it is difficult to determine from the graphic if we applied the rotational boundary conditions. In order to check boundary conditions and get a smooth visualization of the deformation curvatures, the user may resort to remodeling with several element subdivisions within each span.

The stresses in 2-D beam elements consist of a normal stress acting normal to the beam cross section and a transverse shear stress acting on the face of the cross section. The normal stress comes from superposition of the axial stress that is uniform across the section with the bending stress due to the moment on the section. This combination will result in the maximum normal stress occurring either at the top or bottom surface. The transverse shear stress is usually an average across the cross section calculated by the transverse load divided by the area. This obviously does not account for the shear stress variation that occurs across the section from top to bottom [3.3]. The transverse shear stress must be zero at the top and bottom surfaces and has some nonuniform distribution in between that is a function of the cross-section geometry. This variation is usually of minor importance, but the analyst may calculate it if desired.

Most of the available finite programs do not make graphical presentation of the beam stress results. So it reverts to the engineer to evaluate the stress output usually based on values from the printout listing. The engineer also must check for Euler buckling in members that have an axial compressive stress. If the factor of safety against buckling in these members is less than about 4, then the stresses may need correction for the interaction between the axial and flexural stress in that member.

3.4 Case Study

We show a simple beam structure in Figure 3-4 with two different cross sections and two loads. It has simple supports, and we develop a mesh plan in Figure 3-5 with five nodes and four elements. The model input data list is in Table 3-1. The title line is first with the control data line next. Node definition begins with its number, then boundary condition restraints and coordinate locations following. The *z* boundary condition for the 2-D beam element applies to the rotation degree-of-freedom. All the z boundary conditions are free in this model.

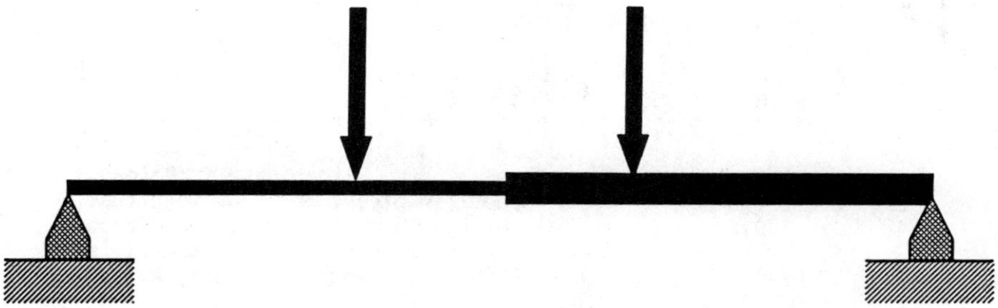

Figure 3-4. Simple Beam with a Cross-Section Change and Two Loads

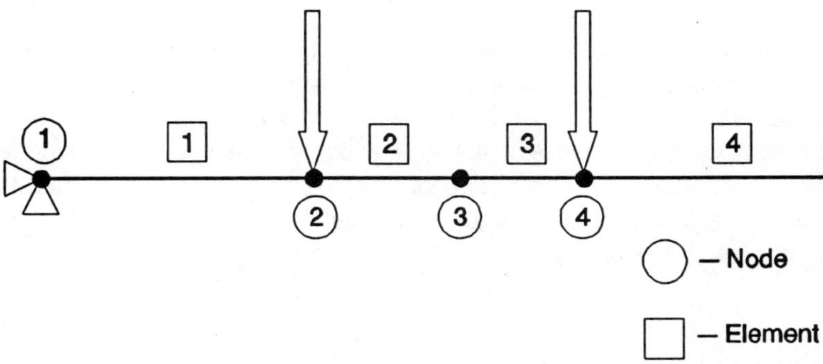

Figure 3-5. Finite Element Model of the Simple Beam

Load specification consists of the node number of application with the load direction and magnitude. The type of element is a beam. The material data is shown with two table entries. Entry 1 has a modulus of elasticity for steel with a cross-section area of 0.4 in^2, moment of inertia of 0.04 in^4, and distance from neutral axis to the surface of 0.5 in. Entry 2 is aluminum with a cross-section area of 0.7 in^2, moment of inertia of 0.06 in^4, and distance from neutral axis of 0.75 in. The element definitions are given by the entry of two node numbers at the endpoints of the element with a material table assignment.

Following execution of the program, the deformed shape plot appears in Figure 3-6, and the results printout is in Table 3-2. In this table the z displacements are the angular rotations of the nodes. The axial stress is

Table 3-1. Input Data File for Beam Model

```
***********|notes inside vertical bars are for data explanation only|
Beam Case Study                                           |title line|
  5    1    1               |5 nodes, 1 element group, 1 load case|
  1    1  1  0      0.000     0.000      0.000          |node number,   |
  2    0  0  0      5.000     0.000      0.000          |x,y,z boundary |
  3    0  0  0      8.000     0.000      0.000          |conditions     |
  4    0  0  0     10.000     0.000      0.000          |0-free, 1-fixed|
  5    0  1  0     15.000     0.000      0.000          |x,y,z coord.   |
  0                         |number of inclined boundary conditions|
  1    2                                        |load case 1, 2 loads|
  2    2  -300.0                 |node 2, y direction, -300 magnitude|
  4    2  -150.0                 |node 4, y direction, -150 magnitude|
  2    4    2             |element type 2-beam, 4 elements, 2 materials|
  1    30.0E+06   0.40   0.04    0.50          |matl 1, E=30E6, A=0.40|
  2    10.0E+07   0.60   0.06    0.75          |        I=0.04, c=0.50|
                                               |matl 2, E=10E6, A=0.60|
                                               |        I=0.06, c=0.75|
  1    1    2    1          |element #, node i, node j, material #|
  2    2    3    1
  3    3    4    2
  4    4    5    2
```

the value from any axial load acting on the element. The flexure stress is due to the resulting bending moment and is the value on the beam top surface when the element definition has nodes I and J arranged left to right. The average shear stress is simply the transverse shear load divided by the cross-section area. The shape of the cross section determines the actual shear stress distribution across the beam height. Finally, a bar graph display of the beam element stresses is given in Figure 3-7.

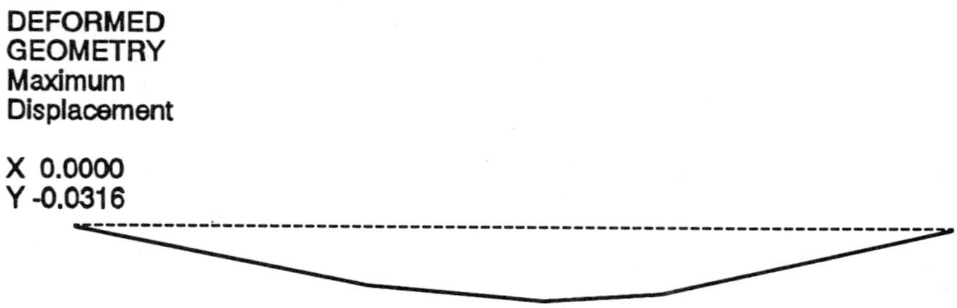

Figure 3-6. Deformed Shape of the Simple Beam

These results show that while the finite element method provides solutions as valid as straight beam theory will allow, there has been no accounting for stress concentration effects where the cross-section change occurred. The designer's job here is to take the results from these

Table 3-2. Results Data File for the Beam Model

D I S P L A C E M E N T S

NODE	X-DISP	Y-DISP	Z-DISP
1	0.000000	0.000000	-0.006026
2	0.000000	-0.025789	-0.003422
3	0.000000	-0.031555	-0.000484
4	0.000000	-0.028968	0.003016
5	0.000000	0.000000	0.007182

S T R E S S E S I N B E A M E L E M E N T
G R O U P 1

ELEM	AXIAL	FLEXURE STRESS		AVG SHEAR
#	STRESS	NODE I	NODE J	STRESS
1	0.	0.	-15625.	625.
2	0.	-15625.	-13750.	-125.
3	0.	-13750.	-12500.	-83.
4	0.	-12500.	0.	-333.

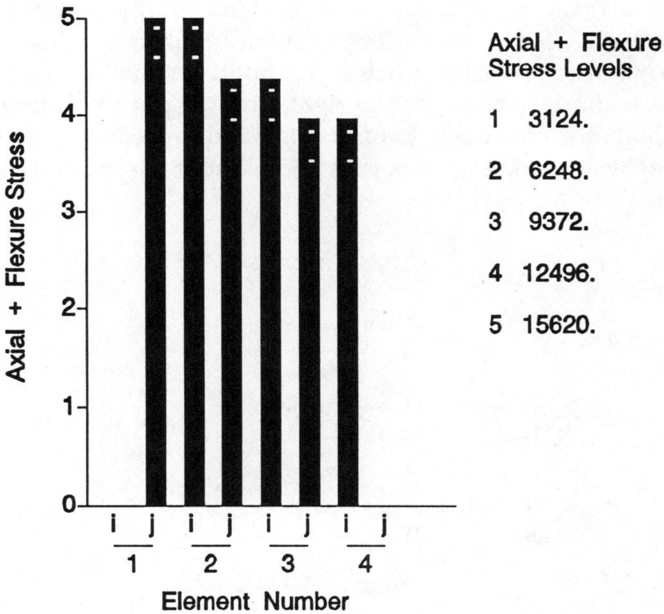

Figure 3-7. Beam Element Stress Display

analyses and then do more detailed modeling of the exact configuration where the cross-section change occurred to evaluate the potential for failure at that location.

3.5 Closure

The use of beam elements in models provides the engineer with the opportunity to solve rather complex beam structures or frameworks that could not easily be done with conventional approaches. It also can include the effects of the stiffness of supporting structures through connection with truss elements or beam elements selected to approximate the support stiffness. In the case of statically indeterminant structures where the supports might have different stiffnesses, the finite element model will provide much better solutions than we can get by conventional approaches. It also can provide the loading that exists at localized areas where cross-section changes or member connections occur. Then we can use the loading in much more detailed models of those regions.

Problems

3.1 The structure shown in Figure P3-1 has a horizontal steel beam welded to a rigid column on the left and simply supported on the right end. There is also a steel rod with pinned attachments to the column and the beam providing support for the beam. The beam cross section is shown on the right, and the rod diameter is 25mm. Evaluate the effectiveness of the steel rod for reducing stress in the beam by analyzing models with and without the rod and comparing results.

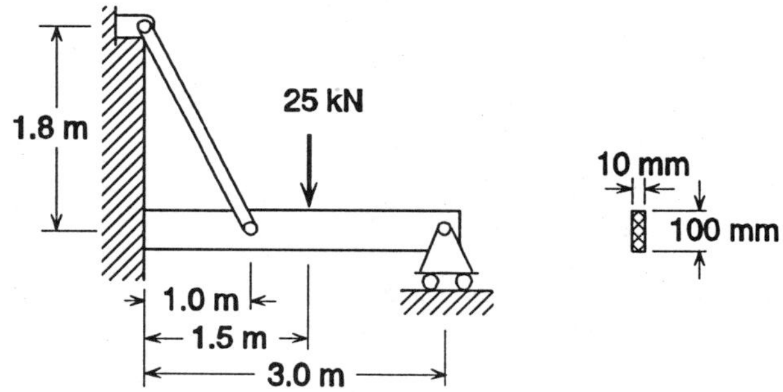

Figure P3-1.

3.2 Curved beams can be approximated by a group of straight beam elements. Opposed diametral forces load the thin circular ring in Figure P3-2. Use symmetry and determine the number of elements required to achieve 5 percent accuracy for the maximum moment and the displacements along and perpendicular to the loaded diameter. Assume that the thickness, $t = r/10$, and that the cross section of the ring is square. The maximum moment and radial displacements are given by

$$M_a = \frac{Pr}{\pi} \qquad \delta_a = \frac{Pr^3}{8EI}\left(\pi - \frac{8}{\pi}\right) \qquad \delta_b = -\frac{Pr^3}{4EI}\left(\frac{4}{\pi} - 1\right)$$

where a is the diameter parallel to the loads, and b is perpendicular to the loads.

3.3 Analyze the bicycle frame design sketched in Figure P3-3. Use a vertical load of 150 lb. at the seat location and 25 lb. at the handlebar location and apply a load factor of 2.5 for inertial loading. Assume for the first analysis that all the members are tubular steel with a 1-in. outside diameter and 0.062-in. wall thickness. From the first analysis, determine if any yield failures are likely if the material is a high-carbon steel with a yield strength of 110 kpsi. If yielding will occur, refine the design by replacement of highly stressed members with a more substantial section or by altering the design layout to eliminate yield failures. If the frame is overdesigned, refine the design to reduce weight. Do the deflections seem excessive? Is there a specific location that seems to be too flexible?

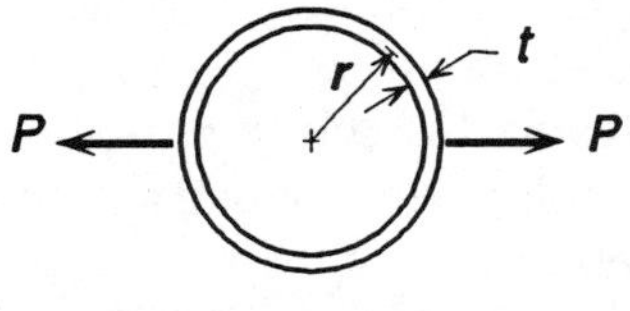

Figure P3-2.

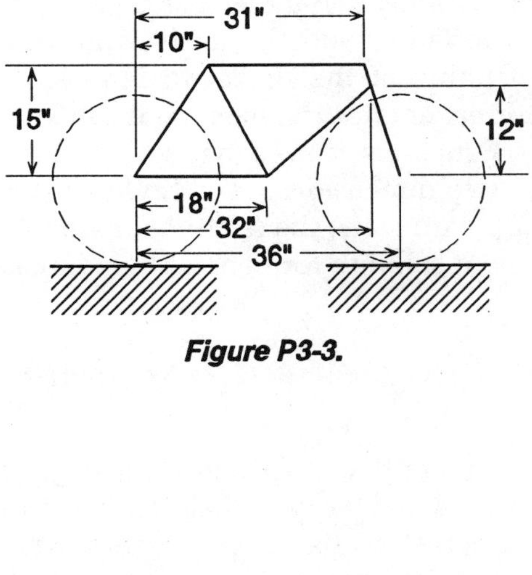

Figure P3-3.

References

3.1 Cook, R. D., Malkus, D. S., and Plesha, M. E., *Concepts and Applications of Finite Element Analysis*, Third Edition, John Wiley and Sons, New York, 1989.

3.2 Logan, D. L., *A First Course in the Finite Element Method*, PWS-KENT Publishing Co., Boston, Massachusetts, 1986

3.3 Cook, R. D. and Young, W. C., *Advanced Mechanics of Materials*, Macmillan Publishing Co., New York, 1985.

CHAPTER 4

TWO-DIMENSIONAL SOLIDS

While the finite element method is very helpful for the solution of truss, beam, and frame problems, the real power of the method shows in application to two- and three-dimensional solid analysis. There are very few closed form solutions to two-dimensional problems, and they are only available for simple geometries and loading conditions. The finite element method, on the other hand, if correctly applied, can provide the solution to most any two-dimensional problem. The correct application is of prime importance, and the analyst makes decisions involving the layout and planning of the model to represent the member under analysis. The correct application must be done to limit solution errors.

Equations derived in theory of elasticity govern the solution to problems in two dimensions. The finite element formulation must satisfy, at least approximately, the relations among displacements, strains, and stresses to find a solution for general two-dimensional problems.

4.1 Element Formulation

There are two shapes of elements used in two-dimensional analysis: the triangle and the quadrilateral. The two basic element shapes may be linear elements or quadratic elements, where linear and quadratic refer to the order of the assumed polynomial displacement interpolation function used within the element area. The linear triangle is the simplest and was the first two-dimensional element developed. Analysts do not use it much now because it requires many more elements to produce a converged and accurate solution compared with the quadrilateral. However, we still examine its formulation both for academic purposes and for its occasional use in coarse to fine mesh transitions for refining models.

The triangular element illustrated in Figure 4-1 defines an area bounded by the three sides connecting three node points. Within the element area the displacement function is assumed to be of the form in equation (4.1),

$$\begin{aligned} u &= a_1 + a_2 x + a_3 y \\ v &= a_4 + a_5 x + a_6 y \end{aligned} \tag{4.1}$$

where u and v are displacement components of a material point within the element field, x and y are coordinates of the point, and a_i, $i=1,2,...6$, are constant coefficients to be determined. This is a linear distribution of the two displacement components for any material point within the element area. The linear function has three undetermined coefficients for each component, and since we have three nodes we may evaluate the three constants by use of the node point values of each component.

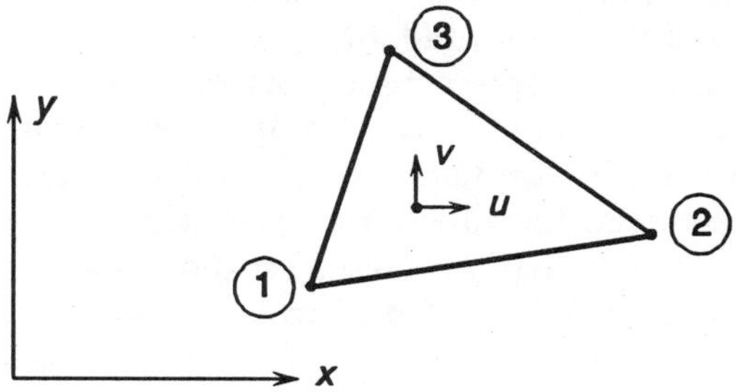

Figure 4-1. Triangular Two-Dimensional Finite Element

Application of the strain-displacement equations to the expressions for u and v illustrates that all three strain components are constant within the element for this assumed displacement field as derived in equation (4.2).

$$\begin{aligned} \epsilon_x &= \frac{\partial u}{\partial x} = a_2 \\ \epsilon_y &= \frac{\partial v}{\partial y} = a_6 \\ \gamma_{xy} &= \frac{\partial u}{\partial y} + \frac{\partial v}{\partial x} = a_3 + a_5 \end{aligned} \tag{4.2}$$

Also, for homogeneous material throughout the element, the stress-strain relations are all constant; therefore, the stress components are also constant.

This displacement formulation also satisfies the compatibility requirements in the theory of elasticity [4.1] for the continuum. The compatibility requirements are that no gaps or overlaps of material may occur during the process of deformation under load. From these equations, we can see that the continuous nature of the function enforces compatibility within the element. On a triangular element, since the interpolation is linear, any edge formed by connecting two nodes that is a straight line before deformation will remain a straight line after deformation. Therefore, any connecting element using the same two nodes for its shared edge satisfies compatibility.

Some of the early finite element programs [4.2] used the triangle element to create a quadrilateral element by subdividing a quadrilateral shape into four triangles using the centroid of the quadrilateral as their apex. After finding the stiffness matrix for each triangle element, assembly of the triangles and condensation of the internal node resulted in the stiffness matrix of the quadrilateral element. This was an effective way to use the triangular element formulation and employ many more elements without tedious input. However, the element of choice now is an isoparametric quadrilateral formulation.

Next we examine the displacement basis for formulation of the isoparametric quadrilateral element. Taig [4.3] developed the element, and Irons [4.4] published its formulation. The quadrilateral element formulation derives from the formulation of a square element. It uses a coordinate system transformation to convert the square to a quadrilateral. Begin with the square element shown in Figure 4-2 with corner nodes.

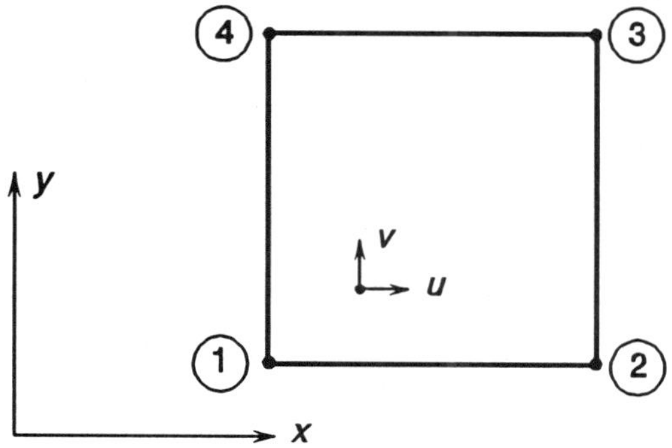

Figure 4-2. Square Two-Dimensional Finite Element

Recognizing that four constants can be evaluated with four nodes, a logical expression for the displacement function components becomes

$$u = a_1 + a_2x + a_3y + a_4xy$$
$$v = a_5 + a_6x + a_7y + a_8xy \, . \tag{4.3}$$

Use of the strain-displacement relations here shows that

$$\epsilon_x = a_2 + a_4y$$
$$\epsilon_y = a_7 + a_8x \tag{4.4}$$
$$\gamma_{xy} = a_3 + a_4x + a_6 + a_8y \, .$$

These equations show that the strain approximation within the element allows ϵ_x to be a linear function of y and ϵ_y to be a linear function of x, while γ_{xy} is a linear function in x and y. So by the addition of one node we have gained a much better approximate solution within the element field for the general case where the strains vary with both x and y throughout the structure domain. Since the stress-strain relations are constant, the stress components may vary similarly within the element field.

Also, in this case the function satisfies compatibility within the element because the function is continuous. Along the element edges for x = constant or y = constant the displacement takes a linear form, and thus remains a straight line between any two of the corner nodes. Therefore element connections to other elements satisfy compatibility as long as corner nodes of one element connect to the corner nodes of the adjacent element. Connection of two adjacent elements to a third element such that the edge of the third element spans two of the adjacent elements' edges, as shown in Figure 4-3, violates compatibility.

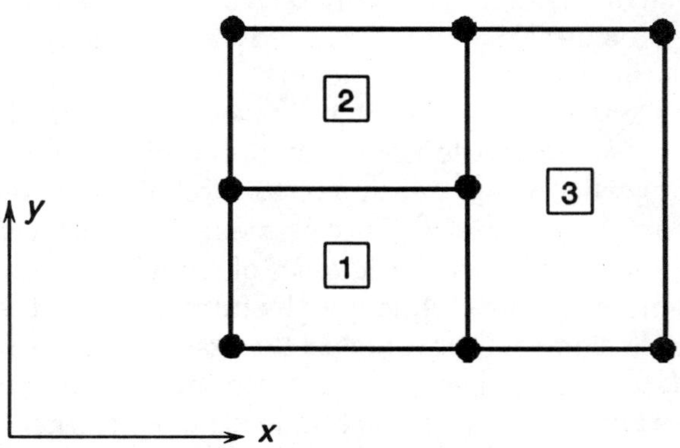

Figure 4-3. Two-Dimensional Element Compatibility Violation

Before continuing the discussion of the square element, make a change of coordinate systems to transform the square into a quadrilateral shape. Square elements can geometrically model very few structures. A coordinate transformation from x,y to ξ,η produces the quadrilateral element sketched in Figure 4-4. We call the element an *isoparametric quadrilateral* because the same interpolation functions (parameters) used to define the displacement field define the geometric transformation.

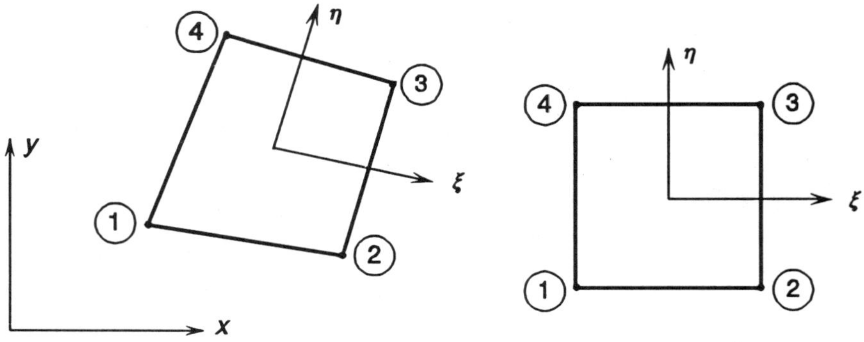

Figure 4-4. Two-Dimensional Quadrilateral Element

4.2 The Finite Element Model

The model plan should begin by choosing the type of element for use. The linear triangle element can easily develop into a mesh inside almost any arbitrary geometry. However, to produce accurate results there must be many of these elements in the model.

In the line element models we have covered thus far, there was little reason for element subdivision other than to define the geometry of the structure. However, two-dimensional cases require element subdivision to achieve an accurate solution. Since element subdivision is required and the exact solution is unknown, a sequence of models with successive mesh refinement is proper. Mesh refinement by further and further subdivision using compatible elements converges to the exact solution. This procedure is known as *h-convergence* because h is a common symbol for step size in numerical operations, and its reduction leads to convergence.

Reaching a refined solution by increasing the order of polynomial approximation within the element is another way to achieve convergence. This has become known as *p-convergence*. A direct conversion of a linear element mesh to quadratic elements will yield a more accurate solution. This is the first step in the *p-convergence* method for numerical convergence on the correct solution. The user may easily use *h-convergence* by successive model building in all finite element programs. However, there are usually only linear and quadratic order elements in the element library of most programs that limit the pursuit of *p-convergence*. There are several new commercial codes becoming available now, and a recent text by Szabo and Babuska [4.5] provides good coverage of the *p-convergence* method theory and application.

In planning the mesh, try to use symmetry whenever possible. The advantages include a reduction of labor of model input, reduction of computer time and cost, and a decrease in computer round-off error in the equation solution because fewer equations exist in the model. There are some drawbacks. Sometimes it becomes more difficult to picture the model. Also, peak stresses may occur along symmetry lines and make it difficult to locate elements properly to show the peak.

Recognize symmetry in two-dimensional objects by observation of geometric patterns that may occur. These may develop by incrementing plane sections, rotating sections about an axis, periodically rotating sections about an axis, or by reflecting a section about a plane. For the symmetric model to provide a solution, the load distribution must also be symmetric on the object. In some cases, we can find solutions for anti-symmetric loading conditions on symmetric objects by proper imposition of displacement boundary conditions.

Displacement boundary conditions enforce symmetry by restricting node points that lie on lines of symmetry to motion along the line of symmetry. For example, look at the simply supported beam with central load in Figure 4-5. It has a vertical plane of symmetry at coordinate $x = 0$.

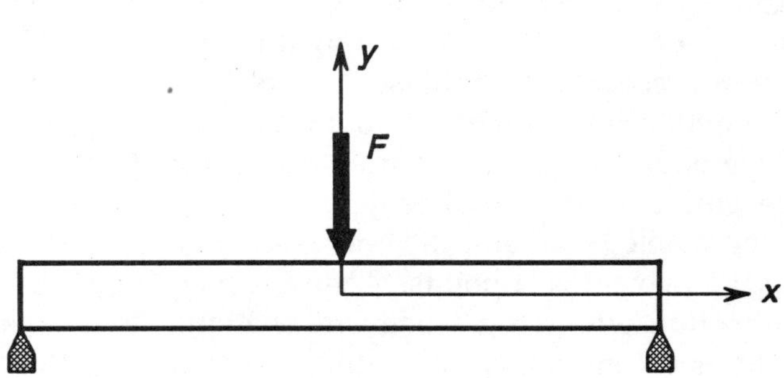

Figure 4-5. Simple Beam with Central Load

When the load applies, the beam will deflect downward and the displacement of every material particle in the right half will be a mirror image of the corresponding particle in the left half. So if the body is symmetric before loading, it is also symmetric after loading. Then we only need to model one half of this beam. If we take the right half, then the outline of the model is in Figure 4-6. The model load reduces by one half because each half of the beam carries its share. The node points that lie on the plane where $x = 0$ are restrained against x-direction motion, but left free to move in the y direction.

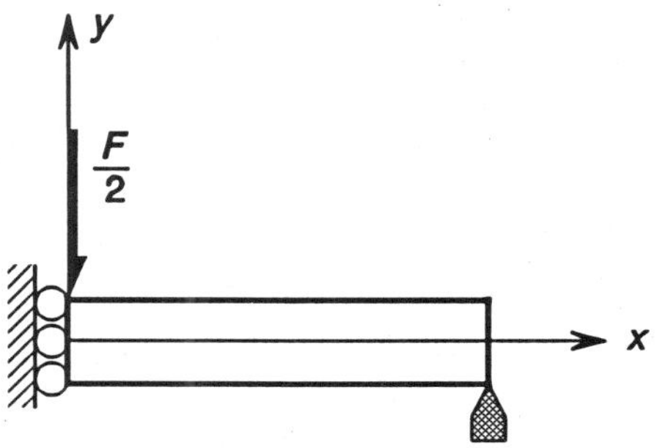

Figure 4-6. One-half Simple Beam Model with Central Load

In a 2-D model, rigid body motion could occur by x translation, y translation, or rotation about the z axis. Examine the displacement restraints applied for symmetry and support conditions, and determine if these restraints will automatically prevent these three rigid body motions. If not, then we need to apply additional boundary conditions to assure that rigid body motion does not occur. If there is a possible rigid body motion and a net external force acting on the body in that direction, then we have not recognized all the support restraints because the body is not in static equilibrium. If there is no net external force, then we may select any single node location for restraint to prevent that motion.

For example, in the simply supported beam in Figure 4-5, assume we model the whole beam and the only restraints are vertical fixed displacement at the two support points. Then the body would be free to move in the x direction with a rigid body translation. To prevent this motion, select any node and apply one x-direction restraint. This one restraint is adequate to prevent rigid body motion. In fact, applying more than one restraint artificially prevents the structure from displacing normally and therefore falsifies the solution. Application of displacement restraints to prevent rigid body motion should not induce any stress conditions in the body.

4.3 Computer Input Assistance

At this point, the analyst should have roughly defined the model. In two-dimensional analyses, we want to develop an adequate subdivision within the area defined by the geometrical boundary. For simple geometries and loadings, typically a regular array of elements will be suitable and are not very difficult to create with simple replication schemes. However, for more complex geometry involving hundreds or thousands of elements, we need the aid of an area or two-dimensional mesh generator. Most programs provide a mesh generation capability in their preprocessor.

One approach for irregular areas is to perform a coordinate transformation mapping from an approximate fit of squares in an integer geometry to the actual physical geometry. This mapping may be done by laying out a rough equivalent of the actual geometry in an integer space where each square in the integer space corresponds to an element. This procedure is illustrated in Figure 4-7.

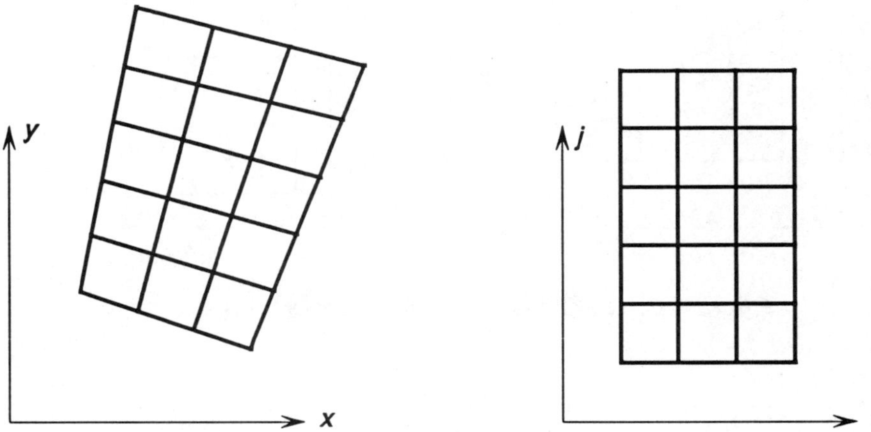

Figure 4-7. Two-Dimensional Mapping Mesh Generation

Sides in the integer geometry correspond to sides in the actual geometry. Sides in the integer geometry must be horizontal or vertical, while sides in the actual geometry can be either straight or curved line segments at any orientation. Correspondence of the closed boundary of the actual geometry with the closed boundary of the integer geometry sets up the mapping. Use of the difference equations, that result from a finite difference approximation of the Laplace equation, determines the interior node locations.

The interior locations result from an iterative solution beginning with a linear interpolation between boundary nodes. Sometimes, in areas of sharply concave boundaries of the actual geometry, linear interpolation of nodes may fall outside the boundary and may not pull back inside the boundary with the iterations. Therefore, the user must examine the generated mesh carefully before proceeding to make sure the mapping was successful. Additional examples of this type of mesh generation are shown in Figure 4-8. In this approach, the user usually has some control over the resulting bandwidth and wavefront by selection of starting location and direction for node and element generation.

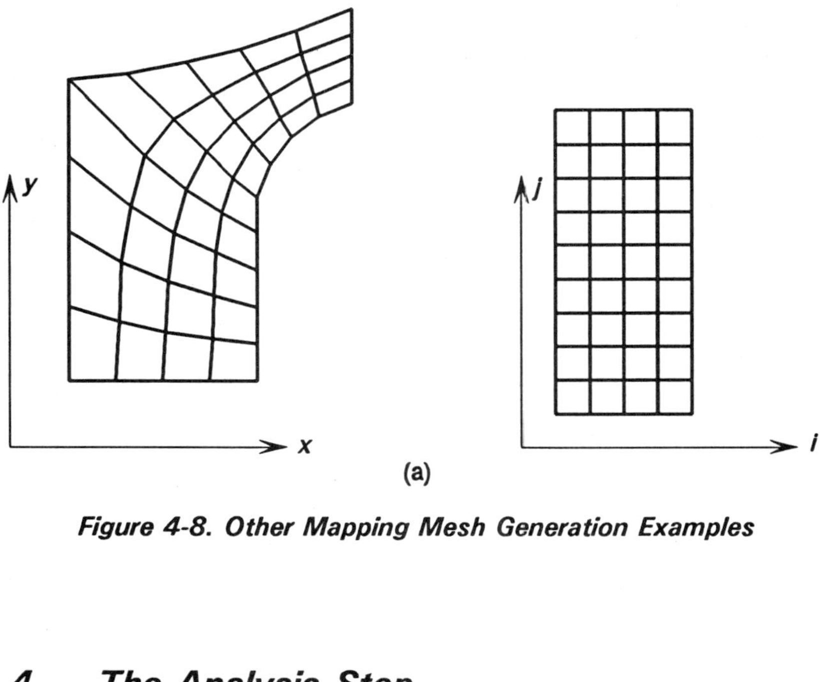

(a)

Figure 4-8. Other Mapping Mesh Generation Examples

4.4 The Analysis Step

In most two-dimensional analyses, there are many more nodes and elements used than with truss, beam, and frame models. Therefore, there is more potential for error in both the analysis execution errors and overall numerical precision errors. If we checked the model thoroughly in the preprocessor, then we should have caught most execution errors. Execution errors arise by not preventing rigid body motion in the model, improperly defining any element, entering incorrect material and physical properties, and many other factors. The error messages presented by the program usually identify these errors rather easily when they occur.

Numerical precision errors may come about through element distortion,

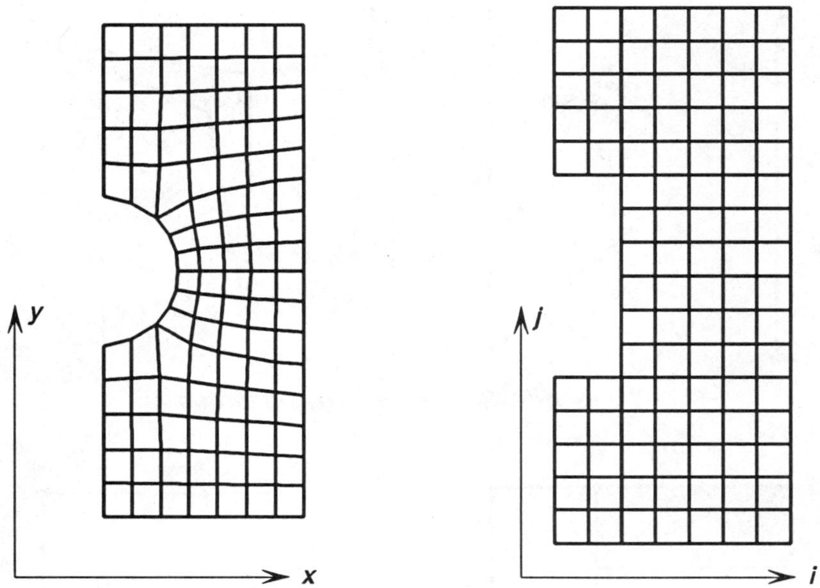

(b)

Figure 4-8. (Continued)

element compatibility violations, and stiffness matrix ill-conditioning caused by large differences in stiffness values of elements [4.6]. Examples of severe element distortion of quadrilateral elements are shown in Figure 4-9. Ideally the elements would remain close to square shape. High aspect ratios, large differences in side length, and very small or large inside angles all contribute to numerical precision errors. In fact, inside angles greater than 180 degrees may cause negative stiffnesses. Most commercial programs will check the element distortion and issue warning messages or cancel execution if the distortion is too high.

Even though most two-dimensional element formulations guarantee satisfaction of the compatibility requirements in theory of elasticity, modeling errors may still violate compatibility. Some of these are illustrated in Figure 4-10.

To satisfy compatibility in corner-noded elements, each element side may only join to a side of one other element. For elements with midside nodes, each must connect to another element at the corners, and all three nodes on a side must connect by use of common nodes.

The solution accuracy in two-dimensional analyses is very dependent on the user's ability to evaluate the results and produce a numerically converged solution. In the truss and beam elements, the element formulation was exact, and therefore there was no concern about interpolation accuracy. However, these two-dimensional elements require

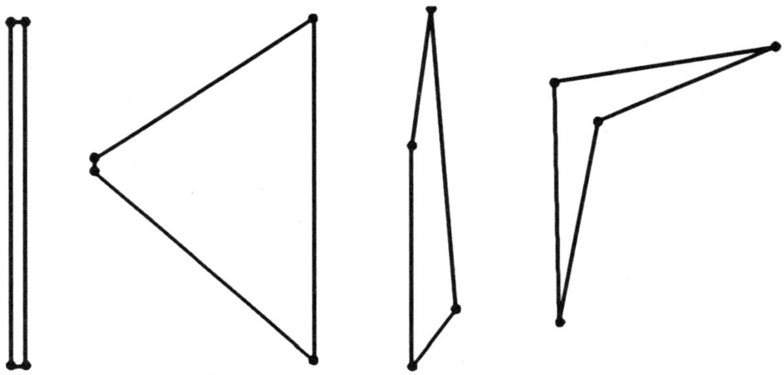

Figure 4-9. Severe Element Distortion

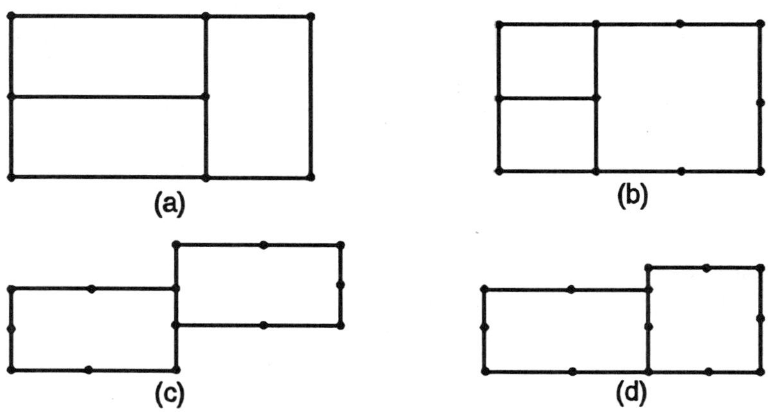

Figure 4-10. Model Compatibility Errors

numerical convergence. We accomplish this convergence through careful evaluation of the output and refinement of the model.

4.5 Output Processing and Evaluation

Completion of the analysis run will produce a listing file and data files for graphic postprocessing. As mentioned before, scan the printout file for errors in interpretation of input data. The data that represent element selection and options, analysis conditions, material and physical properties, and these types of data are relatively easy to scan. Obviously, we cannot easily check the lists of node point and element definitions.

Graphic display of two-dimensional data results is necessary to have any chance at making a complete evaluation. Display graphics will be useful for overall checking of model response as well as location of critical areas. This allows examination of the detailed listing file for specific and accurate values in these critical areas. The deformed shape plots have an exaggeration factor adequate to see the deformation. Check boundary condition enforcement, and make a visual judgment as to whether the deformation agrees with the expected response. In some cases, the exaggeration factor may need to be very high to understand fully how the structure is responding. It may help to visualize the shape you would expect if the material of the actual structure were very soft and easily deformed.

Postprocessors normally present stress results in a contour plot form with a range of stress levels. In examining the plots we can make some simple checks. The boundary conditions of the problem require that any stress component perpendicular to a free surface must be zero. Any stress component perpendicular to a pressure-loaded surface must equal the pressure value. Where a plane of symmetry exists, the stress contour lines should become normal to the symmetry boundary. There should not be any abrupt changes in contour line direction and it should be continuous. The plot should have expected or understandable shape and location of peak values. These plots result from stress component values at the node points. However, in general, the values at a node common to several elements will not be equal. So there must be some manipulation of the computed stresses to produce these contour plots.

Upon examination of the stress contours, the user must make some judgment about the validity of the solution. With only results from one model, we can never be sure that we have a converged or accurate solution. The plan is then to produce multiple models using more refined meshes until the solution has converged. The two-dimensional elements presented in this chapter have mathematically guaranteed convergence to the exact solution as the element size approaches zero. It is then appropriate to run multiple solutions with refined meshes to estimate when the convergence has occurred.

Although the individual stress components may be of interest, the failure criterion should not be the maximum normal stress theory for ductile materials. Its use can result in serious error when the minimum normal stress is of the opposite sign. The maximum shear stress failure criterion is accurate, but be sure to use the true maximum shear stress and not just the maximum in-plane value. The Von Mises equivalent stress based on the distortion energy theory is considered to be the most accurate for ductile materials. If the material is brittle or a composite or some other class of material, then the user must determine what failure criterion is proper to use for the given material.

4.6 Case Study

The case study analyzes a flat bar in tension with a central hole as a typical stress concentration problem. We will analyze this case with the finite element method and compare the results with the theoretical stress concentration factor. The geometry is shown in Figure 4-11. By taking advantage of symmetry, a one-quarter shaded section of the bar defines the model geometry.

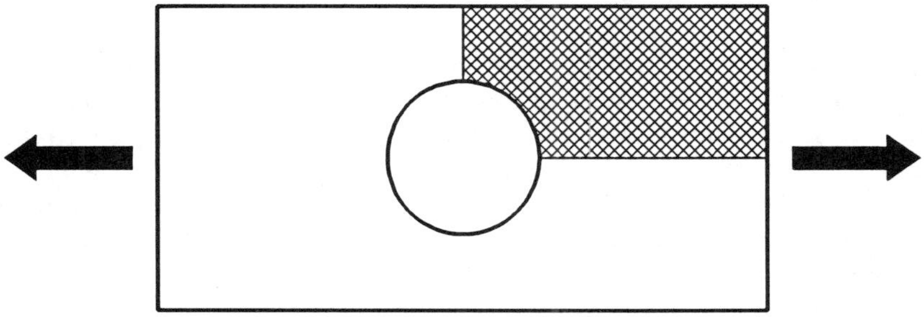

Figure 4-11. Flat Bar with a Central Hole in Tension

The plan for the first finite element model shown in Figure 4-12 has a more refined mesh near the hole because the stress is naturally higher in that area with steeper slopes of change. Displacement restraints apply to the vertical symmetry edge to prevent displacement in the horizontal direction and to the horizontal symmetry edge to prevent displacement in the vertical direction. Node forces calculated and distributed on the right edge provide a uniform stress there of 1 kpsi.

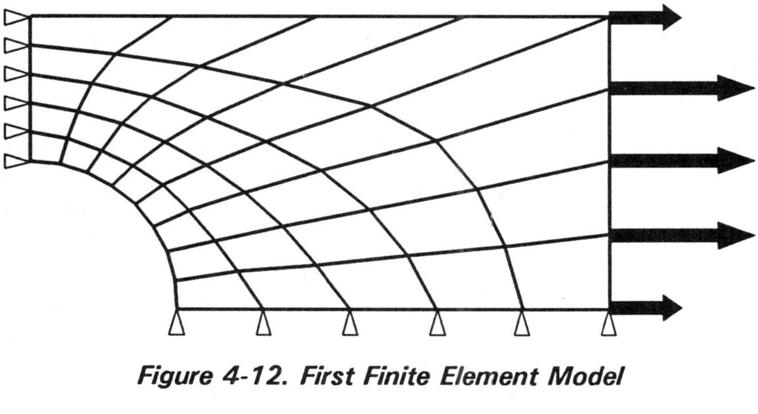

Figure 4-12. First Finite Element Model

After running the analysis, a contour plot of the x-direction stress component appears as shown in Figure 4-13. The contour lines labels correspond to the legend on the left. The element outlines are also shown. The maximum level occurs at the edge of the hole as expected. A clearer view of this region is shown in the enlargement of Figure 4-14. As mentioned before, the finite element method gives an approximate, not an exact, solution. An estimate of the error in the analysis is the range of stress change relative to the average element value across an element. In this case the corner element includes contour levels from 6 to 9 with a total range of about 1500 psi, or about 750 psi from the average. The estimated error is 23 percent.

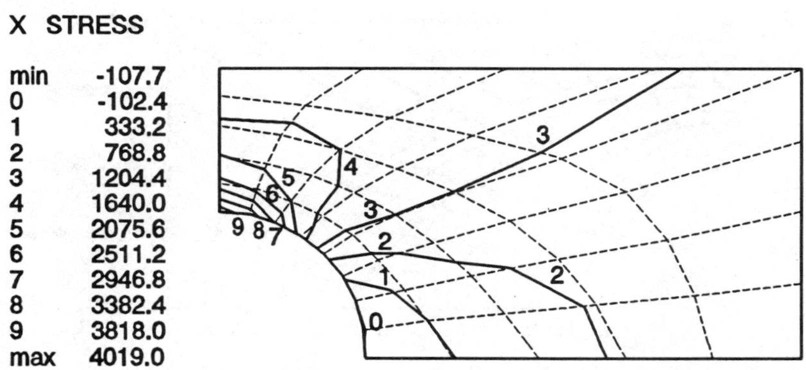

Figure 4-13. Contour Plot of the X Stress Component

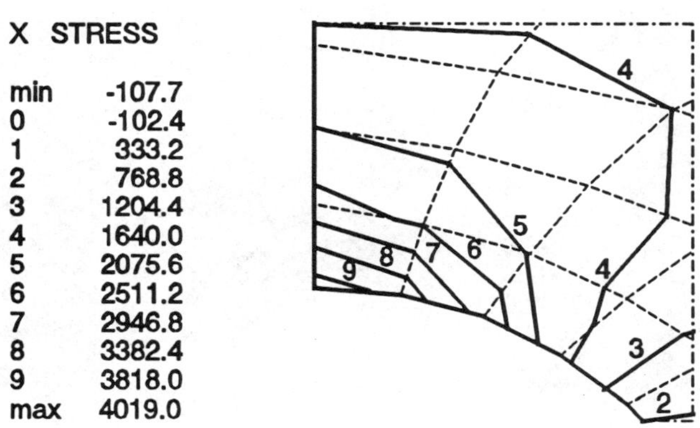

Figure 4-14. Zoom View of the X Stress Contour Plot

This error margin is large, so we should refine the model. A second model produced was still not sufficiently accurate, so we created a third model. This model is shown in Figure 4-15, and it is much more refined around the hole. The zoom view of the x stress component is given in Figure 4-16. We now have a stress range in the corner element of about 700 psi, or about 350 psi from the average, for a 9 percent estimated error. The maximum value at the edge of the hole is 4300 psi. The nominal or average stress on the reduced area section at the hole is 2000 psi which gives a stress concentration factor of 2.15. The theoretical stress concentration factor is 2.18, so the actual error is only -1.4 percent.

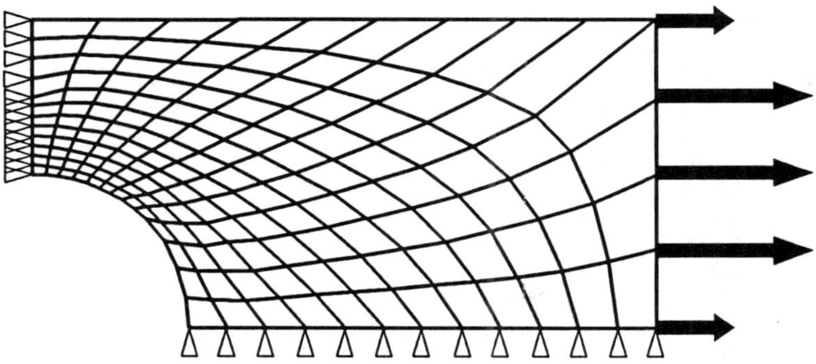

Figure 4-15. Third Finite Element Model

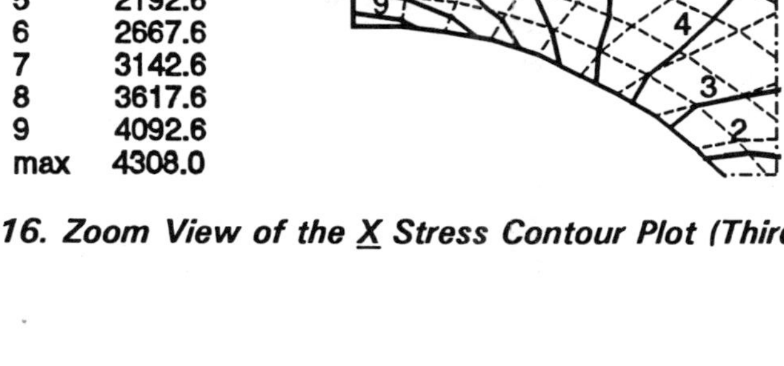

Figure 4-16. Zoom View of the X Stress Contour Plot (Third Model)

4.7 Closure

The application of the finite element method to two- and three-dimensional problems is where its power is really useful. There are very few closed form solutions to these problems, especially for any but the simplest of geometries. As shown by this case study, the engineer can reach a very accurate solution by application of proper techniques and modeling procedures. The accuracy is usually only limited by our willingness to model all the significant features of the problem and pursue the analysis until we reach convergence.

Problems

4.1 Show the convergence of finite element models of a simply supported beam with a uniformly distributed load. Refine meshes to determine the maximum displacement and maximum stress within about 5 percent accuracy. Use a beam length of eight times its height and a unit thickness. Perform the study using each of the following element types if available in the computer program:
(a) linear triangle,
(b) linear quadrilateral,
(c) parabolic triangle, and
(d) parabolic quadrilateral.

Referring to Figure P4-1, the flexural stress and the center deflection including shear deformation are listed in the following equations.

$$\sigma_x = \frac{3qL^2}{4bc^2}$$

$$\delta = \frac{5qL^4}{16Ebc^3} + \frac{3qL^2(1+\nu)}{5Ebc}$$

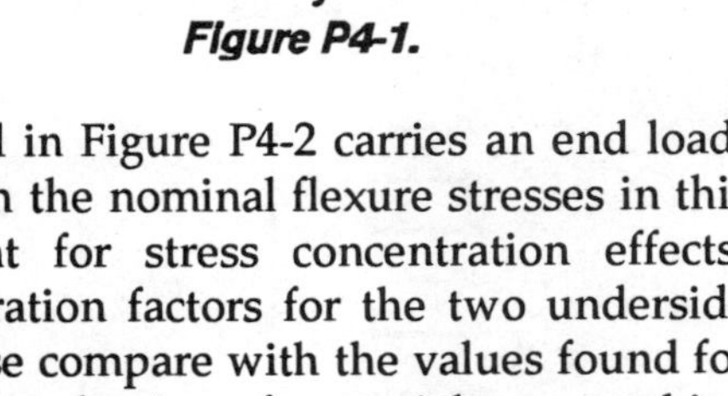

Figure P4-1.

4.2 A machine member illustrated in Figure P4-2 carries an end load. Simple beam equations furnish the nominal flexure stresses in this member but do not account for stress concentration effects. Determine the stress concentration factors for the two underside radius locations. How do these compare with the values found for similar geometries in many mechanics of materials or machine design books?

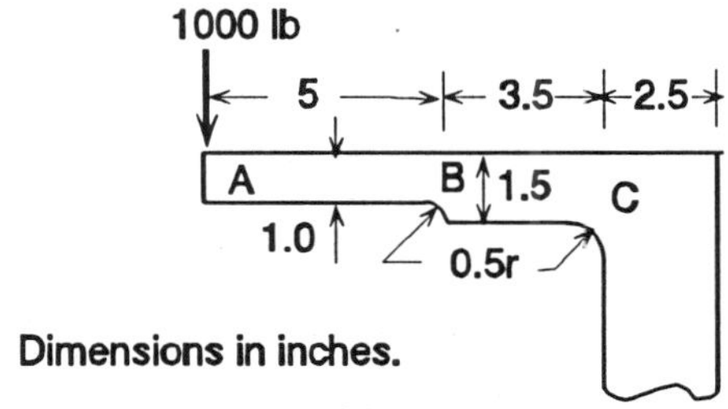

Figure P4-2.

4.3 Analyze the two tensile bar configurations shown in Figure P4-3. Compare your results with published stress concentration factors. Is there a significant interaction effect between the two geometrical discontinuities for the bar in (b)?

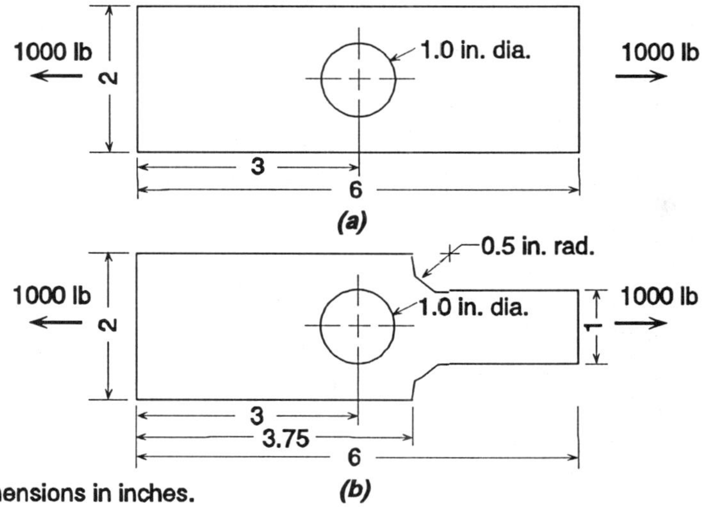

Figure P4-3.

4.4 Analyze the tensile loaded bar with an off-center hole shown in Figure P4-4. Compare results with the closest published results available. Make successive models with the hole moving closer to the side and see if any pattern develops.

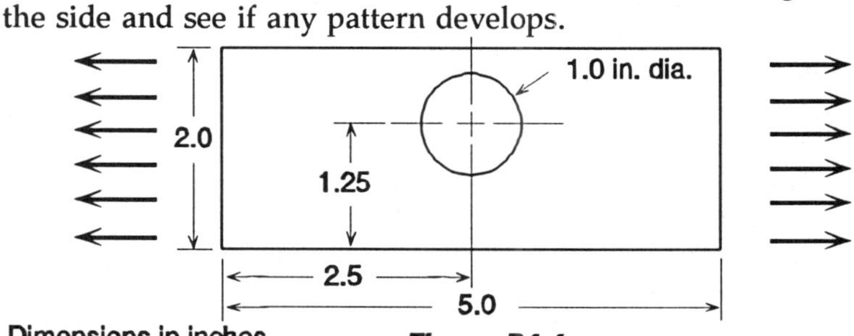

Figure P4-4.

4.5 Determine the stress concentration factor(s) for the notched bar with a center hole in Figure P4-5. Compare with published results for the individual geometries and evaluate any interaction caused by their proximity. You may wish to evaluate additional values of notch radius or varying depths of a constant notch radius.

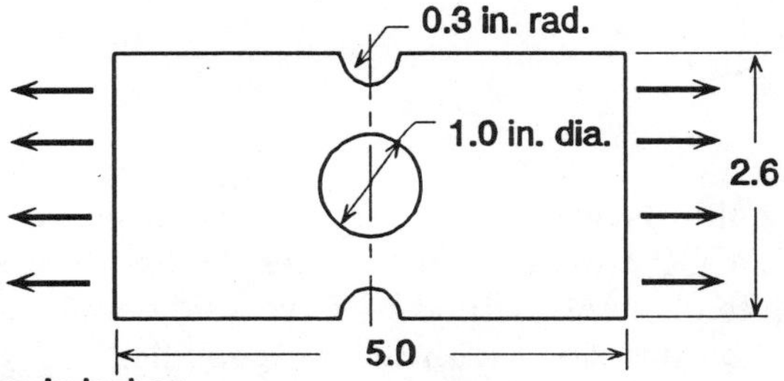

Dimensions in inches. *Figure P4-5.*

References

4.1 Timoshenko, S. and Goodier, J. N., *Theory of Elasticity*, McGraw-Hill, New York, 1951.

4.2 Wilson, E. L. and Jones, R. M., *Finite Element Stress Analysis of Axisymmetric Solids*, Aerospace Corporation, San Bernardino, CA, 1967, Air Force Report No. BSD-TR-67-228, (AD-820991, N.T.I.S.).

4.3 Irons, B. and Ahmad, S., *Techniques of Finite Elements*, Ellis Horwood Limited Publishers, West Sussex, England, 1980.

4.4 Irons, B. M., "Engineering Applications of Numerical Integration in Stiffness Methods," AIAA J., Vol. 4, No. 11, 1966, pp. 2035-2037.

4.5 Szabo, B. and Babuska, I., *Finite Element Analysis*, John Wiley and Sons, New York, 1991.

4.6 Cook, R. D., Malkus, D. S., and Plesha, M. E., *Concepts and Applications of Finite Element Analysis*, Third Edition, John Wiley and Sons, New York, 1989.

APPENDIX

This user's guide is written to describe the operation of the FEPCIP, FEPC, and FEPCOP programs. It assumes the user has been instructed in use of the finite element method to solve stress analysis problems. The programs will be described with their capabilities and general sequence of use for solving a problem. The first-time user should refer to the README file on the program disk for initial setup. In some parts of this guide the user is referred to more complete documentation in the file on the program disk.

These FEPC programs are provided as shareware for educational purposes. They are copyrighted programs and you are not authorized to sell or distribute them for COMMERCIAL purposes. You are free to use, copy, and distribute them for NONCOMMERCIAL uses only if no fee is charged for use, copying, or distribution. Specifically, the programs were designed for use by students in university courses.

If you are an instructor and use the programs in a university course, I would appreciate a simple registration (no fee) of your university, course name or description, and approximate number of students taking the course.

If you use the programs for COMMERCIAL purposes, i.e., employed engineering, consulting, sponsored research, etc., a partial registration fee to continue the software development would be appreciated. For a full registration fee, you will receive the latest version, fully dimensioned, to run on an IBM-PC or compatible with 640K RAM. Please state the current version number of the software you are using. Send inquiries to:

Dr. C. E. Knight
Professor of Mechanical Engineering
914 Ballard Ct.
Blacksburg, VA 24060

The programs are continually under development and your comments concerning present features or future enhancements would be appreciated.

USING FEPC, FEPCIP, AND FEPCOP

FEPC is a program that performs finite element stress analysis of two-dimensional truss, beam, plane solid, or axisymmetric solid structures. There are two companion programs. FEPCIP is the FEPC INPUT PROCESSOR that is used to input and check a model and prepare data files for FEPC. FEPCOP is the FEPC OUTPUT PROCESSOR that reads FEPC output data files and produces graphic displays.

The shareware version of the FEPC programs are currently dimensioned to run in a PC with 256K memory. The dimension limits in the programs are 250 nodes, 250 elements, 10 materials, 100 points, 30 lines, and 20 arcs. Also, the overall model size is limited in FEPC based on number of nodes and average nodal bandwidth. For example, 250 nodes with an average nodal bandwidth of 7, 200 with 10, or 150 with 14 are all maximum model sizes that can be run in FEPC. Dimension limits of the automatic mesh generation grid in FEPCIP are I=15 by J=25. Therefore, a model can be built in FEPCIP even within the 250 node and element limit that is too large to run in FEPC, so plan carefully.

The procedure for solving a problem is to run FEPCIP to create the model, run FEPC to solve the equations, and run FEPCOP to display the results. The FEPCIP program presents the user with menus for interactive input, checking, and storing a model. This creates an analysis file used as input for FEPC. Running FEPC produces a listing file of printout results and files of results used as input for FEPCOP. Graphic displays of deformed shape and stress plots may then be produced by FEPCOP.

ENTERING THE MODEL IN FEPCIP

Before starting to enter the model, develop a node and element numbering plan, boundary conditions, and the load placement for the model. With the FEPCIP.EXE file in the current drive and directory, begin by typing

FEPCIP<CR>

where <CR> means to press the enter or return key. After the FEPCIP logo appears, the program will continue after a short pause.

The screen will clear and the program will automatically detect the proper graphics mode for the supported graphics cards. The supported cards are IBM CGA, EGA, VGA, and MCGA of the PS/2 model 30 along with the Hercules graphics card.

After making the selection the main menu and graphics windows will then appear along with a prompt to SELECT A FUNCTION KEY.

```
INPUT PROCESSOR          FINITE ELEMENT PERSONAL COMPUTER          DATE TIME
 TITLE:
```

F1 FILES F2 MODEL DATA F3 2D AUTOMSH F4 TITLE F6 CLEAR MEM F7 EXIT F8 VIEW OPTS F9 DSPLY OPTS SELECT A FUNCTION KEY	MODEL GRAPHICS WINDOW	MODEL SUMMARY WINDOW

Selection of a menu item by its function key brings up a branch menu for many of the selections.

Key F1 branches to a menu for recalling a previously stored model, storing a new model, or adding a title.

Key F2 branches to a menu for entering or editing all data required for the model.

Key F3 branches to a menu for two-dimensional area mesh generation.

Key F4 prompts the user to input a title for the current model.

Key F6 will clear all the current model data from memory in order to start entering a new model.

Key F7 exits the program.

Key F8 branches to a menu to change the current view of the model.

Key F9 branches to a menu to change the visibility of entities (nodes, elements, loads, etc.) or labels (node numbers, element numbers) on the next redraw of the model.

Every branch menu has a function key selection to return to the previous menu. Many of the selections on the branch menus will branch to additional menus. In each case, following completion of tasks on the current menu, use the previous menu selection to step back through the menus until the modeling is complete.

The general procedure for entering a model is to use the MODEL DATA function key to access the menu for selecting the element type, defining the material properties, defining nodes and elements, setting node displacement restraints, and applying loads. For truss and beam element models, all the model data are entered from this menu and its branch menus.

Two-dimensional solid models using plane stress, plane strain, or axisymmetric elements may first use the 2D AUTOMSH selection to generate the model mesh of nodes and elements. Once the nodes and elements are defined, return to the model data menu to complete the model by material definition, setting node restraints and loads.

The model must be stored on a disk file before exiting the program. The model data may be saved to disk at any time in the progress of building the model. Two files are stored for all complete models under a user-specified filename with file extensions of .MOD and .ANA. If the model is incomplete only the .MOD file is stored and messages denoting the data yet to be defined for the .ANA file are displayed. All the current model data is saved in the .MOD file.

The program operates by using the function keys to select the operation from the menu. When data is required, a prompt appears on the data entry line just below the TITLE: header. The user types in the requested data separated by commas or spaces, followed by the carriage return or enter key.

When the prompt to detect an entity appears on the data entry line, a cursor will appear. If a mouse exists and its driver is loaded, then use the mouse to position the cursor and press the left button to detect or the right button to abort. Otherwise, with no mouse, use the arrow keys to position the cursor at the entity location. Press the space bar to detect, or press the return or enter key to abort the detect and terminate the current operation.

FILES

```
F1 RCL FN.MOD
F2 STO FN.MOD
     & FN.ANA



F10 PREV MENU

SELECT A
FUNCTION KEY
```

Selecting FILES from the main menu branches to the submenu on the left. A previously formed and stored model may be recalled from disk by selecting F1 RCL FN.MOD. The user is prompted to enter the filename, FN, without its .MOD extension. The filename may include the drive designation and path, but it may be a maximum of 20 characters long including the drive designator characters. DO NOT enter any leading blank spaces in the input of the filename. If the file cannot be found, an error message is displayed.

The current model may be stored on disk by selecting F2 STO FN.MOD & FN.ANA. The model currently in memory will be stored in FN.MOD assuming no errors. Also, if the model is complete and ready for analysis, the input file for the FEPC program will be stored in FN.ANA. If the model is incomplete, the user is given messages indicating which data are missing. The store operation may be done at any time during the progressive construction of the model in order to have a place to restart in case of destruction of the current model data in memory. If the files FN.MOD and FN.ANA already exist on the disk, they may be overwritten with the user's consent by the current data in memory each time the store function is executed.

After completing use of this branch menu, select F10 PREV MENU to return to the main menu.

MODEL DATA

Begin entering a new model by selecting F2 MODEL DATA on the FEPCIP main menu. This produces the branch menu shown below.

```
F1 ELEM TYPE
F2 MATL PROP
F3 NODE DEF
F4 ELEM DEF
F5 RESTRAINTS
F6 LOADS

F8 VIEW OPTS
F9 DSPLY OPTS
F10 PREV MENU

SELECT A
FUNCTION KEY
```

If a truss or beam element model is to be entered, then all the model data will be entered from this menu. If a two-dimensional solid is to be entered, then all the data may be entered from this menu, or the 2D AUTO-MSH selection may be used to generate the nodes and elements for the mesh. Once these are generated they may be edited from this menu. If 2D AUTOMSH is to be used it should be done first, or after the element type is selected and material property sets are defined; it will overwrite any existing node and element definitions.

ELEMENT TYPE

The F1 ELEM TYPE selection displays the list of available elements. Use the indicated function key to select the element for the model. Only one element type may be used in a model. After selection the program returns to the previous menu. See the documentation file on the program disk for more detailed instructions.

MATERIAL PROPERTIES

The F2 MATL PROP selection branches to a menu for input and query of material set definition. Up to 10 material property sets may be defined and should be defined in numerical order. Material property sets may include some physical properties depending on the element type. Each element in the model has a material set number associated with it that defines its material properties. If different material or physical properties exist in different parts of the structure, then multiple material sets should be defined before elements are defined so correct assignments may be made at the time of element definition. See the documentation file on the program disk for more detailed instructions.

NODE DEFINITION

The F3 NODE DEF selection branches to a menu to perform node operations. Nodes for truss and beam element models all will be defined in this section. If the 2D automesh option is used for the 2D plane and axisymmetric models, then that should be done first and any additional node operations will be done in this section. Node operations include definition, generating a row of nodes between two defined nodes, moving, deleting, and querying nodes.

A node is defined by its number and coordinate position. A prompt will appear on the data entry line to input a node. Simply enter the node number and its X and Y coordinates. The data must be separated by commas or spaces. All leading blanks are ignored, but do not enter any trailing blanks. If the node is generated properly, it will be displayed on the graphics screen if it lies inside the current window. The starting window is 10 units by 10 units, but it will change automatically if any of the view options are exercised. Autoscale will resize the window so that all currently defined nodes fit inside. The prompt recycles so that the next node may be input. Terminate input by tapping the return or enter key. See the documentation file on the program disk for more detailed instructions.

ELEMENT DEFINITION

The F4 ELEM DEF selection branches to a menu to perform element operations. Elements for truss and beam element models all will be defined in this section. If the 2D automesh option is used for the 2D plane and axisymmetric models, then that should be done first and any additional element operations will be done in this section. Element operations include selecting material, definition, generating a row of elements from a starting element, modifying, deleting, and querying elements.

Single elements are defined by user selection of nodes for each element. The user is prompted to detect each node needed for the element definition. The order of node selection on two node elements is of no consequence. ***The nodes for four node elements must be picked in a counterclockwise order surrounding the element area.*** Elements are numbered in numerical order as they are defined. Their material set assigned is the current material set. Each element is drawn on the graphics screen as it is defined. The node prompts recycle to define the next element, and will continue until the right mouse button is pressed or the return or enter key is pressed at the node detect prompt to terminate element definition. See the documentation file on the program disk for more detailed instructions.

RESTRAINTS

The F5 RESTRAINTS selection is for applying node displacement boundary conditions. By default all nodes displacement components are free to take on nonzero values appropriate to the structure response under load. The components that must be zero for the model to behave properly are specified to be fixed. The menus that appear allow the user to set values for the restraints and then pick the nodes to which the set values apply. See the documentation file on the program disk for more detailed instructions.

LOADS

The F6 LOADS selection is for applying loads to the model. Loads may be node forces or element edge pressures (for 2-D solid elements). The menus that appear for node forces allow the user to set component values for the loads and then pick the nodes to which the set values apply. Edge pressure is applied by input of the pressure value and then selection of the elements to which the set pressure applies. Menu selections also allow deletion or query of forces and pressures. See the documentation file on the program disk for more detailed instructions.

2-D AUTOMESH GENERATION

This section of the program is used for area mesh generation of two-dimensional plane stress, plane strain, or axisymmetric models. The principle of the approach is a mapping of an integer area grid into the geometrical area of the model. The geometrical area is defined using point locations, lines, and arcs. The perimeter of the geometrical area is defined by the complete set of lines and arcs that enclose the area.

The integer area grid will have lines that correspond to the lines and arcs of the geometrical area. Plan the correspondence by imagining or physically sketching on square grid paper the perimeter in the integer area. Use integer coordinates I and J with a range of 1 to IMAX and 1 to JMAX, respectively. IMAX and JMAX values are listed in the first section of this guide. A 1-by-1 square in the integer area grid will map to an element in the geometrical area. Grid points in the integer area grid will map to node points in the geometrical area model.

Lines in the integer area can only be lines of constant I or lines of constant J. The perimeter must be defined by a head-to-tail connection of lines in a counterclockwise(ccw) direction around the area. The length of line in the integer area is equal to the number of elements desired along the corresponding line or arc in the geometrical area.

The process involves defining the geometrical points needed to describe the model area, then defining lines or arcs using those points that complete the model perimeter. Next, plan the corresponding integer area grid to be mapped into the geometry of the model.

After all the geometric points, lines, and arcs have been entered, area mesh generation can begin. The genmesh function presents a prompt to pick the starting point of the area. This point on the geometry will correspond to the 1,1 I,J coordinate location on the integer area. A series of prompts then proceeds for the detection of a line or arc, the number of elements on that line or arc, and the direction of the corresponding line in the integer I,J area.

The first line or arc detected must have the selected starting point as one of its endpoints. The next line or arc picked must have the other endpoint of the first line or arc as one of its endpoints. Each successive line or arc picked must then connect to the other endpoint of the previous line or arc. This continues until the perimeter of the geometry is closed and the endpoint of the last line in the integer area must be back at the starting point; i.e., the perimeter in the geometry area and the perimeter in the integer area must close simultaneously. Both of these perimeters must progress ccw around the area.

When a line or arc is picked, the entry of number of elements determines the length of the line in the integer area. The direction entry chooses one of four allowable line directions in the integer area. The directions are labeled 1, 2, 3, and 4, which correspond to right(+I), up(+J), left(-I), and down(-J), respectively, in the I,J coordinates.

The integer area of the model must lie in the positive quadrant of I,J coordinates. Since the starting point in the integer area is at 1,1, and the perimeter must be ccw, the direction for the first line or arc must be 1. The direction for the second line or arc picked may be 1 or 2. Successive lines may have any direction values as long as some lines with directions 1 and 2 are used before any with directions 3 or 4, so that the I and J coordinate values always remain positive. The total number of elements on all lines in the 1 direction must match the total number in the 3 direction, and the total number in the 2 direction must match the number in the 4 direction.

The bandwidth of the structure stiffness matrix is minimized by making the number of elements in the I direction smaller than in the J direction. The limits are IMAX-1 elements in the I direction and JMAX-1 elements in the J direction. However, no model may have more than the maximum number of nodes or elements listed in the first section of this guide.

Mapping is an iterative process of distorting the integer area to fit in the geometry area. After a few iterations a mesh will be drawn on the screen. If it appears to be suitable then it can be accepted, or more iterations may be requested to make it smoother. If it is unacceptable then a different integer area may be tried.

The menu of functions for mesh generation is

```
F1 POINT
F2 LINE
F3 ARC
F4 GENMESH


F8 VIEW OPTS
F9 DSPLY OPTS
F10 PREV MENU

SELECT A
FUNCTION KEY
```

POINT

Points are used to define lines and arcs that make up the model's geometric perimeter. Two points are needed to define a line, and three points along the arc are needed to define an arc. Points are input by their coordinate location. Selection of key F1 POINT brings up a submenu for creating, modifying, or deleting points. See the documentation file on the program disk for more detailed instructions.

LINE

A straight line may be used to represent all or part of any straight edge on the model. More than one line on an edge might be used to produce different element spacings along the edge. If two or more lines are used on any single edge, they should be connected in series with no overlap. Selection of key F2 LINE produces a submenu for creating, modifying, or deleting lines. A line is defined by picking two points at the ends of the line. See the documentation file on the program disk for more detailed instructions.

ARC

An arc may be used to represent all or part of any circular arc on the model up to 180 degrees included angle. If more than one arc is used on a circular arc of the model, then they should be connected in series. Three points along the arc are needed for the definition. They are the two endpoints and an intermediate point. Another point is created during definition of the arc at the arc's center of curvature. This may cause the autoscale function to reduce the model scale substantially if the arc radius is very large, in order to fit all the points on the graphics screen. Selection of key F3 ARC produces a submenu for creating, modifying, and deleting arcs. See the documentation file on the program disk for more detailed instructions.

GENERATE MESH

Selection of the F4 GENMESH function begins a series of prompts and inputs to define the meshing area. If the element type has not been selected, the element menu will be presented for a choice. If more than one material set has been defined, then the prompt to enter material set number will appear. Enter the set number to be assigned to all the elements defined using the mesh generator.

Following these conditional entries the prompt to detect the start point appears. This is a geometric point on the model that corresponds to the 1,1 point in the I,J integer area. Next, the prompt to detect a line or arc begins the sequence of perimeter definition. Detect the line or arc that connects to the starting point and starts on the ccw path around the perimeter by positioning the cursor on the line or arc center and pressing the spacebar or the right mouse button. Following detection, enter the number of elements along the line or arc. Then enter the direction number of the line in the integer area (1, 2, 3, or 4). The set of prompts to detect line or arc, enter number of elements, and enter direction all cycle until the user terminates input by entry of the return key.

The user should be sure the geometry perimeter is closed before terminating. The program checks that the integer area is closed, and if so begins iterating on the mapping. This may take a few minutes. If the integer area is not closed, a program message reports this condition and the geometry is redrawn. Another trial to input the perimeter may begin with selection of the automesh function.

If the integer area did close, after a few iterations the mesh is drawn on the screen with the prompt for more mesh iteration (Y or N). If it needs additional smoothing, enter Y. More iterations will then be done and the prompt will reappear. If it looks acceptable or is to be redone differently, then enter N.

The next question asks if it's OK to keep (Y OR N). Enter Y to keep the mesh, or enter N to discard this mesh and redo the GENMESH function with another plan.

Once an acceptable mesh of nodes and elements is kept, return to the MODEL DATA menu to apply the displacement boundary conditions and loads, and perhaps define a material property set.

When the model data is complete, go to the files menu and store the model and analysis files to disk. This also should be done periodically during building of the model in case of unexpected termination of the session or a different path chosen to complete the model.

VIEW OPTIONS

This selection appears on many of the branch menus to allow exercising the view options without retracing the menus to reach them. The following menu appears on selection of view opts.

```
F1 AUTOSCALE
F2 ZOOM
F3 MAGNIFY
F4 CENTER



F10 PREV MENU

SELECT A
FUNCTION KEY
```

AUTOSCALE

Selection of autoscale automatically scales the graphics window to include all currently defined points and nodes.

ZOOM

This function allows the user to select a portion of the graphics window which is then scaled to fit the full window. The user is prompted to detect the two corners of the zoom area.

MAGNIFY

This option changes the size of the model displayed. The user is prompted to enter the magnification factor. A positive value must be entered; values larger than one will increase the size of the drawing and values smaller than one will decrease the size.

Subsequent use of the magnify command will enlarge (or decrease) the model display with respect to its current size. For example, magnifying your model by two and then by three produces an image six times larger than the original.

CENTER

The model may be moved by selecting a new center of the graphics window. The user is prompted to locate the new center.

DISPLAY OPTIONS

Display options control which entities and labels are visible when a graphics plot is done. A branch menu appears.

```
F1 ENTITY SW
F2 LABEL SW

F4 MONO/COLOR



F10 PREV MENU

SELECT A
FUNCTION KEY
```

ENTITY SWITCH

An entity switch setting is off or on to control the individual entity's visibility. Selection of this function brings up a submenu listing all the entities and prompting the user to select one.

Selection of a function key will produce a prompt to change its current setting by default. A Y or return key entry will accept the prompt question.

LABEL SWITCH

This function controls the display of labels (numerals) for nodes and elements on the graphic model. A submenu allows selection of a function key producing a prompt to change its current setting by default. A Y or return key entry will accept the default.

MONO/COLOR

This function switches the display between black and white or color when the computer has a VGA graphics board. On other graphics boards the display is always black and white. Switching the VGA to black and white allows the screen graphics to be dumped to a black-and-white printer without the loss of character intensity that sometimes happens in such screen dumping. Since only the drawing color palette is changed with this switch, the change occurs when the next drawing is done after the switch. Execute the function key again to return to a color display.

THE ANALYSIS BY FEPC

When a model has been developed and saved, it is complete and ready to be processed by the finite element processor, FEPC. After exit from FEPCIP, and before starting FEPC, be sure that the filename.ANA file can be accessed by FEPC. Copy it to the FEPC diskette in the same directory where FEPC.EXE resides, or use the drive and path designation in the filename.

RUNNING FEPC

With the FEPC.EXE file in the current drive and directory, begin by typing

FEPC<CR>

After the FEPC logo appears, a prompt will appear to enter the model filename (20 characters max), with drive and/or path designation but without the .ANA extension.

As the computations proceed, messages will appear on the screen reporting the computation step in progress. If errors occur, error messages will also appear on the screen. FEPC creates some other files as it runs. There is a listing file of all the printed output labeled filename.LST. This file should be studied by the user after an analysis to check the input data interpreted by FEPC and all the numerical output. A file labeled filename.MSH stores the node and element data for FEPCOP. A file labeled filename.NVL stores the node displacement and element stress data for FEPCOP.

Some other files are also created during the FEPC run; these are deleted upon normal termination of the program so the disk that stores the .ANA file must have some excess space for these files during runtime. If the run terminates abnormally some of these files may still be on the disk with extensions of .ELM and .LOD. These and other output files will be overwritten when running a model with the same filename.

If the FEPC run was successful, then FEPCOP may be used to display the results in graphic form. If the run was not successful, then examine the filename.LST file for data errors or error messages that may help to correct the model.

The output from a FEPC run is stored in a listing file called filename.LST, where the filename is the same as the model file name entered when beginning the FEPC analysis. This file includes a listing of all the input data as well as all the numerical results. See the documentation file on the program disk for more detailed instructions.

FEPC ERROR MESSAGES

1 - OUT OF SPACE, MODEL IS TOO LARGE (I)

The model is too large to run in FEPC. Reduce the model size in FEPCIP and try again. Consult the program limits given before.

2 - NODE 'n' HAS BEEN PLACED ON AN INCLINED BOUNDARY, BUT IT IS ALREADY CONSTRAINED AGAINST X OR Y DISP. OR BOTH

An inclined boundary angle is specified for node n, but an x or y restraint was also specified which is incompatible. Edit the model in FEPCIP.

3 - FEPC.EXE file not found

The FEPC.EXE file must reside in the current drive and directory to execute.

4 - 5 No longer used.

6 - 'm' ELEM IS HIGHER THAN NO. OF ELEMENTS IN THE GROUP

The filename.ANA file has been corrupted because element number m is higher than the total number of elements. The *.ANA file is an ASCII file so it may be printed or edited. Examine its contents in comparison with the data printout in the filename.LST file.

7 - ELEMENT NO 1 IS NOT DEFINED FIRST

The filename.ANA file has been corrupted because element number 1 is not defined first in the list of element definitions. The *.ANA file is an ASCII file so it may be printed or edited. Examine its contents in comparison with the data printout in the filename.LST file.

8 - YOU HAVE A ZERO LENGTH ELEMENT #'m'

A beam or truss element is defined using the same node for both ends or the two nodes defining the element have coincident coordinate locations.

9 - BAD ELEMENT #'m'

A quadrilateral element is improperly defined or is too distorted. Check for cw node order definition around the element (it should be ccw), inside angles between sides greater than 180 degrees, butterfly-shaped element, or a triangle formed by using one node for two corners (this is legal if the last two nodes in the element definition are the same).

10 - STIFFNESS MATRIX NOT POSITIVE DEFINITE, NEGATIVE STIFFNESS DIAGONAL TERM FOR EQUATION 'n', VALUE = '#'

During solution of the system equations, a negative diagonal term is found which means that the equations cannot be solved. The equation number corresponds to the free node degree-of-freedom in the system ordered consecutively with node numbers. These are listed in the filename.LST file produced in the FEPC run. Find the node number from this list, then examine the elements defined using this node number for errors. If the equation number is 1 or the last equation number, then the error is probably due to lack of sufficient displacement restraints to prevent rigid body motion.

11 - INCLINED BOUNDARY ANGLE MUST BE BETWEEN -89.99 AND +89.99 DEGR

The inclined boundary angle input is outside the allowable range.

GRAPHIC RESULTS USING FEPCOP

After a successful run by FEPC, results files, filename.MSH and filename.NVL, will have been created on the disk. These are the input files for output processing by FEPCOP.

With the FEPCOP.EXE file in the current drive and directory, begin by typing

FEPCOP<CR>

where <CR> means to press the enter or return key. After the FEPCOP logo appears, the program continues after a short pause.

The screen will clear and a prompt will appear to

ENTER MODEL FILE NAME (NO EXT) -

Enter the filename (20 characters max), with drive and/or path designation, but without the .MSH or .NVL extension. Some messages will appear noting the progress of calculations, and then the screen will clear and the FEPCOP MAIN MENU is displayed along with a prompt to SELECT A FUNCTION KEY.

```
F1 DEFORMED
F2 X-STRESS
F3 Y-STRESS
F4 XY-STRESS
F5 T-STRESS
F6 VON MISES
F7 TRUSS STRS
F8 BEAM STRS
F9 OPTIONS
F10 EXIT


SELECT A
FUNCTION KEY
```

Selection of F1 DEFORMED brings up a branch menu. Selection of F1 PLOT in this branch produces a deformed shape plot of the element mesh superimposed over the undeformed model. This plot shows the finite element mesh when the node displacements are scaled and added to the node coordinates so that the deformed shape is exaggerated. In truss and beam models the deformed mesh is superimposed over the undeformed mesh plot. In 2-D solid models the deformed mesh is superimposed over the outer boundary of the undeformed shape. The displacement scale factor may be changed in the OPTIONS menu to increase or decrease the plotted deformation. An additional submenu appears for view options of the plot.

Selection of F2 ANIMATE in the branch menu produces a sequential mesh plot of truss and beam models or a boundary outline plot of 2-D models, showing the progressive deformation as the load is applied cyclicly. Press any key to terminate the animation.

The next five function key selections on the main menu show the stress contour plots developed in 2-D solid models for the indicated components of stress. Each function is accompanied by a legend of the contour values and the view options menu. The T-stress component is nonzero only for the axisymmetric element models and represents the hoop stress in the axisymmetric structure. The Von Mises equivalent stress is calculated based on the distortion energy failure theorem using all the stress components calculated in the loaded model.

TRUSS STRS

This function is used to display the results in truss element models. The user may select a plot of the axial force or stress in all truss elements. The plot is in a bar chart format with the heights scaled to the maximum value in any element. Plus or minus signs are drawn on the bar near the top to indicate whether the member is in tension or compression.

BEAM STRS

This function is used to display the results in beam element models. The user may select to plot the axial, flexure, average transverse shear, or the maximum combined axial plus flexure stress. These are also in bar chart format with signs indicated near the top of each bar. The axial and transverse shear stresses are constant along an element length, so one bar per element is sufficient. The flexure stress component varies linearly along the element length, so a bar is plotted for the value at each end. Two bars are also plotted for the combined axial plus flexure stress. The sign of the combined stress is the same as the sign of the axial stress which is the combination producing the largest magnitude.

OPTIONS

```
F1 NODE SW
F2 ELEM SW
F3 DISP SCALE

F10 PREV MENU

SELECT A
FUNCTION KEY
```

Function key F9 OPTIONS produces a submenu. Selecting F1 adds node symbols to the 2-D stress plots, and F2 adds element outlines inside the 2-D boundary for stress plots. Both of these selections produce a prompt to switch the current setting. Answering Y or by default a return key entry will make the change. Selection F3 allows the scale factor for the deformed shape plots to be changed by prompting for a new scale factor, with the current scale factor shown as the default value. Enter a larger value to increase the exaggeration or a smaller value to decrease it.

INDEX